RATIONALITY, RELATIVISM AND THE HUMAN SCIENCES

STUDIES OF THE
GREATER PHILADELPHIA PHILOSOPHY CONSORTIUM

J. MARGOLIS, M. KRAUSZ and R.M. BURIAN
Series Editors

J. Margolis, M. Krausz and R.M. Burian, eds. *Rationality, Relativism and the Human Sciences.* 1986. ISBN 90-247-3271-9

RATIONALITY, RELATIVISM AND THE HUMAN SCIENCES

edited by

J. MARGOLIS
Temple University

M. KRAUSZ
Bryn Mawr College

R.M. BURIAN
Virginia Polytechnic Institute and State University

1986 **MARTINUS NIJHOFF PUBLISHERS**
a member of the KLUWER ACADEMIC PUBLISHERS GROUP
DORDRECHT / BOSTON / LANCASTER

Distributors

for the United States and Canada: Kluwer Academic Publishers, 101 Philip Drive, Assinippi Park, Norwell, MA 02061, USA
for the UK and Ireland: Kluwer Academic Publishers, MTP Press Limited, Falcon House, Queen Square, Lancaster LA1 1RN, UK
for all other countries: Kluwer Academic Publishers Group, Distribution Center, P.O. Box 322, 3300 AH Dordrecht, The Netherlands

Library of Congress Cataloging in Publication Data

Rationality, relativism, and the human sciences.

(Studies of the Greater Philadelphia Philosophy Consortium)
Includes bibliographical references and index.
1. Philosophy--Addresses, essays, lectures.
I. Margolis, Joseph Zalman, 1924- . II. Krausz, Michael. III. Burian, Richard M. IV. Series.
B29.R34 1986 190 85-26003
ISBN 90-247-3271-9

ISBN 90-247-3271-9 (this volume)

PRINTED IN THE NETHERLANDS

CONTENTS

IV

PREFACE

The Greater Philadelphia Philosophy Consortium was launched in the early eighties. It began during a particularly lean period in the American economy. But its success is linked as much to the need to be in touch with the rapidly changing currents of the philosophical climate as with the need to insure an adequately stocked professional community in the Philadelphia area faced, perhaps permanently, with the threat of increasing attrition. The member schools of the Consortium now include Bryn Mawr College, the University of Pennsylvania, Temple University, and Villanova University, that is, the schools of the area that offer advanced degrees in philosophy. The philosophy faculties of these schools form the core of the Consortium, which offers graduate students the instructional and library facilities of each member school. The Consortium is also supported by the associated faculties of other regional schools that do not offer advanced degrees – notably, those at Drexel University, Haverford College, La Salle University, and Swarthmore College – both philosophers and members of other departments as well as interested and professionally qualified persons from the entire region. The affiliated and core professionals now number several hundreds, and the Consortium's various ventures have been received most enthusiastically by the academic community.

At this moment, the Consortium is planning its fifth year of what it calls the Conferences on the Philosophy of the Human Studies. These Conferences were begun sometime before the Consortium was formed, through the cooperation of two of the editors of the present collection, representing Bryn Mawr and Temple. The Conferences – and the whole spirit of the Consortium – have been focused on the need to bridge the gap between Anglo-American and Continental movements in philosophy and to widen the professional focus of philosophy itself so that it is genuinely interdisciplinary and informed at the highest level about pioneer work in all the other sciences and humanities, with special attention to the developing role of what have come to be called the human studies or human sciences.

The essays here collected are the first published efforts of this large cooperative venture. The papers have been culled from some of the recent Conferences as well as from the ongoing work of the fortnightly sessions of the Consortium's own study group, that has gradually achieved an enviable level of professional exchange. These papers, then, suggest some of the new directions in which the professionals of our area are pressing, both in terms of a developing but generously construed regional consensus and in terms of coming to terms in a fresh way with the principal, most influential currents of contemporary philosophy.

We have not wished to hurry into print, and we have in fact waited rather patiently for a naturally cohering collection of papers to gather before launching any publications. Now, however, that this maiden volume is actually a public event, we invite the philosophical community to enter into the spirit and letter of exchange that, within the Consortium, we have, to our great delight, found it possible to sustain. In fact, the cooperative nature of the present volume says as much about the willingness of our visiting lecturers to enter into our venture as it does about a very active local membership.

Philadelphia, Pennsylvania
June 1985

The Editors

Part One

WHY STUDIES OF HUMAN CAPACITIES MODELED ON IDEAL NATURAL SCIENCE CAN NEVER ACHIEVE THEIR GOAL

HUBERT L. DREYFUS

1. Introduction[1]

Ever since the human sciences (anthropology, economics, linguistics, political science, psychology, and sociology) rather slowly and tentatively separated themselves from philosophy they have had to explain why the break was never clean, why they have a massive amount of data but little theoretical understanding, in short, why they have not found their own secure way of doing science as the natural sciences have. A recent discussion of this subject has taken the form of a three way debate in which Charles Taylor defends a sharp distinction between the kind of interpretive understanding we can expect in the sciences of man and the kind of theoretical explanation we have achieved in the sciences of nature; Richard Rorty claims there is no essential difference between the *Geistes-* and *Naturwissenschaften*, only between normal and non-normal science; and I draw on ethnomethodology to argue that the human sciences which try to emulate the physical sciences cannot be normal because they must but cannot account for their own conditions of possibility.[2]

After reading Michel Foucault, I now think that my attempt to distinguish human from natural science on the basis of the different way their background practices are internal to each fails to pinpoint their essential difference. What Foucault calls the sciences of *man*, are, indeed, defined in such a way as to make their conditions of possibility internal to them. This raises insurmountable difficulties for these Kantian disciplines but one can always substitute for the sciences of *man* the scientific study of *human beings*, which I will call the human sciences. Still, the strikingly different status of the natural and human sciences is too pervasive and constant to be merely an historical accident. To characterize and account for this difference I need to explain why Taylor, as understood by Rorty, and the ethnomethodologists, as understood by me, fail to make their case that the human sciences cannot be stable and cumulative, i.e., normal like the natural sciences. I will then use some of their

Margolis, J., Krausz, M. and Burian, R.M., (eds.), Rationality, Relativism and the Human Sciences.
 ISBN 90-247-3271-9.

insights plus Heidegger's account of the essential structure of the everyday world and of theory to try once more to locate the line between the *Geistes-* and *Naturwissenschaften*. I will argue that, although one cannot *prove* that the human sciences which try to emulate the physical sciences can never be closed and predictive and thus not normal sciences like the physical sciences, one can give arguments which lead one to expect them to fall short of the ideal of natural science in just the way they do. I will then suggest a way of studying human beings that can be rigorous and repeatable, even normal, without attempting to approach the ideal aimed at by theoretical natural science.

It is generally accepted that natural science (especially physical science) is normally *stable* and *cumulative*. For extended periods, sciences like physics exhibit an agreed upon way of doing research. Occasionally disagreement arises as to how to account for anomalies and about what counts as evidence and valid arguments. This constitutes a crisis which continues until the anomalies are removed by some new scheme which gains agreement, re-establishing normal science. On this view, the natural sciences are not as constant nor as cumulative as once believed, but there is still stability and progress.

The human sciences, on the other hand, have never been stable or cumulative. They suffer from what Foucault calls "essential instability."[3] These "dubious disciplines," to use Foucault's pejorative phrase, do not progress through revolutions like physical science, but merely go through episodes in which certain fads tend to dominate research until some competing fad lures most researchers onto its bandwagon. One style of research gives way to another not because the new research is based upon a theory which explains certain anomalies the old theory failed to explain, but because researchers have become bored and discouraged with the old approach. The new style allows everyone to forget the old questions by introducing challenging new methods and problems. Thus most of the human sciences are in a state of constant factionalism and reorganization, but this is not a crisis – a period of competing paradigms between periods of normal science. It is a pre-paradigm state. As a result, the human sciences are not even normally stable nor are they even relatively cumulative.

In his article, "Interpretation and the Sciences of Man,"[4] Charles Taylor sets out to show that there is an essential difference between the subject matter of the natural and social sciences that explains why social sciences which seek to imitate the predictive power of the physical sciences cannot become normal. According to Taylor the essential difference between the natural and the social sciences is that the natural sciences study mere objects, whereas the human sciences study self-interpreting beings and so have to take their object's changing self-interpretations into account. As Mark Okrent puts it succinctly: "For this position the human sciences are radically distinct from the natural sciences. Their 'object' (what is investigated) is understanding, not things, their aim is understanding, not explanation, and their method is interpretative, not empirical."[5]

To take Taylor's example, recent political scientists assume that the political domain is essentially restricted to the adjudication of individual preferences. This definition of politics makes possible theories based on polls to determine preferences. But this definition is just one possible interpretation of the political domain. It can change with changing social self-interpretations. The point is *not* that the actual techniques used to collect data can be criticized as not sufficiently controlled, etc. (such a critique could also be appropriate in physics), nor that the assumptions guiding the collection of data might be false and therefore lead to false data (this can happen in any science since all data are theory laden). Taylor's example is meant to show, rather, that *interpretation* both on the part of the researchers and on the part of the groups they study, is involved in determining what are to count as political facts. Taylor claims that similar conflicts of interpretation concerning the domain of investigation need not occur in the natural sciences, since the objects they study are not self-interpreting.

Rorty points out in response to Taylor that conflicts of interpretation are not restricted to, nor inevitable in, the human sciences. Although at present what counts as political is disputed within political science itself while the physicists take for granted some shared notion of what counts as nature, leaving the question of how to interpret the paradoxes of quantum reality to the philosophers, this has not always been the case. There is no in-principle reason why it might not come to pass that all political scientists would once again come to agree on the ontology of the political domain as they did in the classical age, and physicists again disagree about the basic categories of nature as they did when quantum theory was first proposed. Then, the background interpretations and practices in the human sciences which define their domains, at present a constant issue within the disciplines themselves, would be taken for granted by all practioners, while the background interpretations of the physicists would again become problematic.

According to Rorty, then, the difference Taylor finds between the *Geistes-* and *Naturwissenschaften*, viz. that in the human sciences the background interpretation establishing the regional ontology that defines the domain is an issue for the discipline while in the nature sciences it is not, is actually the difference between both pre-paradigm and revolutionary science, on the one hand, and normal science on the other. Rorty concludes that the only important distinction between kinds of sciences is the distinction between normal and non-normal science – a distinction which cuts orthogonally across the old *Geistes/Naturwissenschaften* boundaries.

> The line that Taylor is describing is not the line between the human and the non-human but between that portion of the field of inquiry where we feel rather uncertain that we have the right vocabulary at hand and that portion where we feel certain that we do. This *does*, at the moment, roughly coincide with the distinction between the fields of the *Geistes-* and *Naturwissenschaften*. But this coincidence may be *mere* coincidence.[6]

Thus Rorty holds that *both* the natural and human sciences are capable of periods of relative predictive power, stability and cumulative development, as well as periods of confusion and failure. While, at present, the human sciences are unable to find an agreed vocabulary, this fact has nothing to do with the essential nature of man or of science. It is just an accident of the current scene. Taylor, on the other hand, insists that the success of modern science shows that we have finally got the right vocabulary for the natural world, and that when we see that the natural world is essentially meaningless and the human world is essentially meaningful and self-interpreting, we will see why we can never find the right vocabulary in the human sciences.

I am not convinced by either party. Rorty is right that in a successful science of human behavior human self-interpretation would turn out to be irrelevant, just as the sociobiologists, for example, hope. Still I agree with Taylor that the instability of the human sciences is no mere coincidence, but has something to do with the essential characteristics of their subject matter. There must be something about human beings, on the one hand, and the nature of theoretical explanation, on the other, which do not fit together. This alone would explain why the human sciences have never become normal.

I will try to defend what I take to be the truth in Taylor's position in Part IV, but first we must consider an interesting recent approach to the human sciences which argues that the problems of the human sciences are not merely accidental as Rorty claims nor are they due, as Taylor thinks, to the ontological status of their subject matter. Rather, the problems of the human sciences arise from the way their subject matter, *man*, has been defined.

2. Foucault's claim that the sciences of man cannot be stable because of the way man is defined

In *The Order of Things: An Archaeology of the Human Sciences*, Michel Foucault argues that the human sciences "are doomed to an essential instability," because of the way they have been set up.[7] As Foucault puts it:

> What explains the difficulty of the 'human sciences,' their precariousness, their uncertainty as sciences, their dangerous familiarity with philosophy

> . . . is not, as is often stated, the extreme density of their object; it is not the metaphysical status or the inerasable transcendence of this man they speak of, but rather the complexity of the epistemological configuration in which they find themselves placed . . ..[8]

According to Foucault the human sciences involve a unique human self-interpretation, which reaches its fullest expression in Kant. They interpret their domain of investigation, man as a transcendental/empirical double. Man is both a meaning giver or interpreter who constitutes the world and determines what counts as objects, and yet at the same time man is an object in the world like any other. This conception of man makes human self-interpretation essential to an understanding of human beings while at the same time stipulating that human beings are meaningless objects amenable to the sort of theory characteristic of the natural sciences.

From Kant to Donald Davidson philosophers have accepted this definition of man. In various ways they have argued that man can be viewed under two aspects, i.e. that one can give both a totally intentionalistic and a totally physicalistic account of all human activities. Such a view runs into many problems, e.g., it has led both to affirming and denying free will. But even if this ''double aspect'' theory could be made to work, the sciences of man which operate within this metaphysical perspective would still necessarily be abnormal. Since each aspect is supposed to yield a complete description of man without reference to the other, in any discipline in the human sciences one would expect an interpretive and a materialist school to confront each other, each claiming to have the right vocabulary in which there was no place for the vocabulary of the other. If the materialistic side could complete its research program and produce a predictive theory, it would no doubt carry the day, since all scientists would like to be materialists, but for reasons which I will develop in Section IV, materialism, currently understood as mechanism, cannot account for human phenomena. Thus the sciences of man as defined by Kant inevitably exhibit the ''wearisome'' form noted by Foucault. An interpretative social scientist like Taylor will claim that there is something about human subjects or meaningful practices which make them essentially different from objects, and so intrinsically incapable of being brought under objective theory, while the other side will argue that human activity is identical with physical motion and so must fall under some sort of objective laws. Each side then supports its own approach by convincingly showing the inadequacy and incompleteness of the other.

Foucault's approach allows us to understand the force and the fallaciousness of what seems to me the only interesting argument other than Taylor's that the human sciences can never be scientific, viz., the ethnomethodologists' claim that the background skills presupposed by social scientists (not the self-interpretations of the human beings they study) are internal to the human sciences but cannot be treated theoretically.[9]

To understand this challenge to objective social science, we must first distinguish what we will call "ontological" from what we will call "ontic" background practices. The practices which determined what counts as language, history, the political, etc., embody an ontology, and restablish a general domain and so can be called *ontological.* To take Taylor's example: American social practices involve, and most American political scientists uncritically accept, an interpretation of political life as negotiation among private individuals each with his or her own preferences. *Ontic* practices, on the other hand, enable the practitioners in a discipline to collect data by picking out and isolating specific facts within an already established ontological domain. Such skills enable coroners, poll takers, linguists, psychologists, etc., to produce the data social scientists attempt to relate by rules or laws. Thus we can dinstinguish the *constitutive* or *ontological* background practices shared by a society and the sciences which study it from the *methodological* or *ontic* background practices used by the scientists studying various aspects of that society.

Ontological practices involve an interpretation of what constitutes a social domain – an interpretation which is subject to challenge by a counter-interpretation and can result in social revolution, a scientific revolution or merely in a pre-paradigm change of bandwagon. *Ontic* practices, however, do not involve an interpretation in the same sense. They are rather techniques for isolating and establishing facts within an already taken for granted domain. To take suicide statistics as an example, what is at stake when Garfinkel, in his doctoral dissertation, criticizes Durkheim's use of suicide statistics is not a conflict of interpretations as to what counts as a suicide, but as an attempt to spell out the specific considerations experts use to determine that a death is a suicide. The point of the interpretive social scientists is simply that suicides are not given as brute facts but must be produced by a set of specific practices. In general, ethnomethodologists want to study "the objective reality of social facts as an ongoing accomplishment."[10]

Garfinkel has recently stressed that not only social scientists but also physicists, astronomers, etc., have elaborate shared skills which enable them to produce objective data. Garfinkel thinks that these background practices should be a part of physics just as the sociologists' background practices should be a part of sociology. Unfortunately, natural scientists turn out to have what Garfinkel considers a reprehensible "amnesia" concerning their background skills when they write up their results in publishable aricles.

I think that Garfinkel misunderstands this amnesia. What he has discovered is that scientists in a normal science have a double skill. They have a first order skill for producing decontextualized data by using instruments which register only context-independent features such as color, weight, electric change, rate of diffusion, etc., for only such decontextualized facts can fill in the variables over which scientific laws range. But, normal scientists also have learned a further skill by which they eliminate all reference to their taken-for-granted, shared, first-order skills from their final reports. Thus a description of the practices by which everyday objects get worked over into theory laden physical facts is external to normal natural science. Descriptions of such skills appear and should appear in the journals of the ethnomethodologists, not in *The Physical Review*.

The ethnomethodologists would prefer to see the natural scientists themselves reporting on their background skills, if only in the interest of honesty and completeness. What interests me, however, is the ethnomethodologists' implicit stronger claim that the human sciences *must* include an account of their own fact producing practices, not just because otherwise they might produce faulty data – that is true in any science – but because, as I once put it when I held this view, the data-producing practices of human scientists are an essential part of human activity and so cannot be omitted from human science.

The question thus arises: Can the human sciences provide a scientific account of their own background practices? At this point, the interpretive social scientists imply, objective social science breaks down. If background skills could be captured in strict rules then we could have objective social science, e.g. "cognitive" sociology, but skills are so context-specific no law-like or rule-like generalization can be made about them. As I prefer to put it, scientific skills must be taught by apprenticeship and practice and can at best be formulated in *ceteris paribus* rules whose everything-else-being-equal clause presupposes a shared background of practices which cannot be further specified in rules.

Although I no longer accept this way of arguing for the limits of the human sciences, I think that the claim that skills cannot be treated theoretically is correct and very important. Since I wish to draw on this point later, I will pause here to sketch the phenomenological considerations which lead me to conclude that, however skilled behavior is produced by the nervous system, there is no reason to think that it can be accounted for as the law- or rule-governed organization of elementary movements or actions i.e., no reason to claim that there can be a behaviorist or cognitivist theory of skills.

Skills may, and, indeed, usually do begin as rules for manipulating context-free elements. This is the element of truth in cognitivism. Thus a chess beginner must follow strict rules relating such features as center control, material balance, etc. After one begins to understand a domain, however, one sees meaningful aspects not context-free features. Thus a more experienced chess player sees context-dependent aspects, like unbalanced pawn structure or weakness on the king side. A further stage of proficiency is achieved when, after a great deal of experience, one is able to see a situation as having a certain significance tending towards a certain outcome and certain aspects of the situation standing out as salient in relation to that end. Finally, after even more experience the current situation is perceived as similar to another already experienced one and one sees immediately what is required. The chessmaster, for example, sees the issues in a position almost immediately, and the right move just pops into his head. *There is no reason to suppose the beginner's features and rules, or any other features and rules, play any role in such expert performance.*[11] Thus the social scientists' skills which make possible the isolation of features and rules can no more be understood theoretically, i.e., in terms of features and rules, than any other skills, so if the social sciences must have a theory of their own background practices, then, so the argument goes, the social science will always be incomplete and unstable.

I hope this argument sounds convincing, since I was until recently convinced. Foucault, however, attempts to show both why it is convincing and also why it does not have the significance one might suppose. In *The Order of Things*, he claims that the demand that each human science produce a theoretical account of that science's foundations is a natural outcome of trying to have a science of *man*. Since man is understood as both the meaningful activity which determines what counts as facts, and at the same time an object totally amenable to theoretical investigation, the sciences of man, in trying to subsume every aspect of man, must keep trying to objectify the background practices which provide their objective data:

> [T]he human sciences . . . find themselves treating as their object what is in fact their condition of possibility. They are always animated, therefore, by a sort of transcendental mobility . . ..[12]

If Foucault is right, however, problems of accounting for the meaningful background practices of the human sciences which seem to put an in-principle limit on the scientific study of human beings, are only problems for a science that studies a peculiar human self-interpretation, the transcendental/empirical double Foucault calls *man*. In spite of Taylor's and Garfinkel's arguments it would seem that neither ontological nor ontic background practices put an in-principle barrier in the path of the scientific study of *human beings*. Only those sciences in the grip of the transcendental/empirical double are compelled either to objectify everything, even the background practices which make them possible, or else to give up theory and turn to interpretation. In *The Order of Things*, Foucault, like Rorty, holds that the instability of the sciences of *man* is in no way essential to, or even relevant to, the scientific study of *human beings*. Just as Rorty thinks that current cognitive science might well be on the right track, Foucault, at the time he wrote his critique of the sciences of man, felt that an objective, stable science of human beings was about to be achieved by structuralists such as Jacques Lacan, Noam Chomsky and Claude Lévi-Strauss.[13]

I have argued elsewhere that cognitivism and structuralism have fared no better than their predecessors.[14] If this is so, it seems clear that the failure of the sciences which seek a science of human beings on the model of physics cannot be accounted for by showing the contradictions in one particular historical configuration. Granted the inevitable failure of the sciences of *man*, we must still ask: What are the obstacles in the way of *any* theoretical study of human capacities modeled on ideal physical theory?[15]

3. The argument that there cannot be a theory of human capacities based on elements abstracted from the everyday context of human activity

Rorty allows that the human sciences have been in trouble from Dilthey to behaviorism, and Foucault believes he explains why, but neither can explain why cognitivism and structuralism (which are supposed to have superseded hermeneutics, behaviorism and the sciences of *man*) are no more predictive and so no more normal than their predecessors. I will now argue that this non-normality is inevitable given the relation of ideal theory to human activity. Inevitable difficulties arise for the human sciences not from the definition of man but in so far as the current sciences of human beings imitate physical science in seeking a theory that precisely predicts events in the everyday world using context-free features abstracted from that world.

First I will have to define what I mean by theory and by science. Theorizing is a special form of intellectual activity, discovered by Socrates and refined by the philosophical tradition. Ideal theory has six essential characteristics, never fully achieved, but approached to varying degrees. The first three are introduced by Socrates. (1) *Explicitness.* Ideally a theory should not be based on intuition and interpretation but should be spelled out so completely that it can be understood by any rational being. (2) *Universality.* Theory should hold true for all places at all times. (3) *Abstractedness.* A theory must not require reference to particular examples. In the *Euthyphro* Socrates introduces these requirements when he assumes that normal behavior must be based on abstract, universal principles and so asks the prophet Euthyphro to justify his behavior by providing an explicit, universal and abstract definition of piety – angrily rejecting Euthyphro's appeal to examples and his own special intuition.

Descartes and Kant complete the Socratic account of theory by adding two more requirements. (4) *Discreteness.* A theory must be stated in terms of context-free elements – elements which make no reference to human interests, traditions, institutions, etc. (5) *Systematicity*. A theory must be a new whole in which decontextualized elements (attributes, features, factors, etc.) are related to each other by rules or laws.[16]

Plato clairvoyantly expressed all five characteristics in the myth of the cave: the theorist must remove his object of knowledge from the everyday, perceptual, social world in order to see the universal relations between the explicit and abstract elements, in this case the ideas. Freed from all context, the elements form a system of their own – all ideas are organized by the idea of the Good. Plato saw that while everyday understanding is implicit, concrete, local, holistic and partial, theories, by contrast, are explicit, abstract, universal, and range over elements organized into a new total whole.

In the *Republic* Plato put the natural sciences of his day on only the third level of his divided line since they did not satisfy two further conditions: indubitability and rational grounding. Nonetheless, the theoretical explanation of nature developed by modern natural science, although still only on Plato's third level, is the most successful version of the sort of theory Plato envisaged. Indeed, while relinquishing indubitable and grounded knowledge, the modern natural sciences have been able to add a sixth characteristic of ideal theory. It is this characteristic and its advantages which is sought in vain by the social sciences.

(6) *Prediction and completeness*. The description of the domain investigated must be complete, i.e. it must specify all types of changes of elements which effect the domain and must specify their effects. In natural sciences such as physics this completeness permits precise prediction.

For this discussion it is irrelevant whether natural scientists ever achieve Plato's ideal of theoretical understanding or whether, as is now generally believed, shared interpretations, metaphors, specific examples, etc. play a crucial role in the activities of normal scientists and their understanding and acceptance of theories. All I need claim for my argument is that theories, as they approach the ideal, make no essential reference to shared background interpretations, metaphors, examples, etc. It is also irrelevant for my argument whether scientists ever succeed in formulating ideal theories, i.e. whether they ever find the basic elements their theories require or whether they ever succeed in isolating the systems they study so that their explanations are truly complete and predictive.

The human sciences rightly admire the theoretical power of the natural sciences and seek to emulate them. The behaviorist seeks a complete description of human action in terms of elements of behavior and proposes covering laws to predict the occurrence of particular behavioral events; the cognitivist assumes human beings are devices whose functional components are in discrete states and seeks to explain this device's capacities by analyzing the capacities of its components and the rule-like relations holding between the states. Both approaches aim, in the ideal limit, at attaining the sort of complete explanatory account which allows one to predict precisely the effects of alterations in an object's internal and external environment.

To be scientific in the strong sense defined above, the sciences which study human capacities as their object must begin by defining an appropriate level of investigation. They cannot seek to predict all the motions of human bodies. For example, they do not wish to predict people's coughs or the trajectory of their falls. Rather, they pick out as their object specifically human activities such as speaking, political action, social interaction, etc., and seek to give a theoretical explanation of this everyday activity following the model of natural science, i.e. by subsuming elements abstracted from these everyday events under laws or rules. This is, indeed, the only available approach if one seeks a science modeled on idealized physics. However, I will try to show that such a project must fail because (1) the features necessary for a theory of everyday capacities cannot be totally abstracted from the everyday background practices, and, (2) these background practices cannot themselves be analyzed in terms of elements related by laws or rules.

The natural sciences abandon everyday meaning to achieve precise predictions. Their decontextualized elements and laws are meaningless, and so are the phenomena they are meant to explain. Natural science, according to Heidegger is: "[A] subspecies [of knowledge] which has the legitimate task of grasping the present-at-hand [decontextualized elements] in its essential unintelligibility."[17] Just as physical science predicts and explains everyday changes of place in terms of meaningless, context-independent properties such as mass and position which can be abstracted from the everyday world, so a theoretical science of human beings seeking to emulate natural science would have to abstract meaningless, context-free features or attributes from everyday, context-dependent, meaningful human activities, and then predict and explain these everyday activities in terms of formal relations between these elements. So, to take Foucault's favorite examples, Chomsky looks for syntactic elements and formal transformational rules to account for everyday judgments of grammaticality, and Lévi-Strauss abstracts objects used in exchanges and then formalizes their role in everyday social interactions.

I will now argue that this approach, which succeeds brilliantly in the natural sciences, cannot work in the human sciences. Insofar as these would-be sciences follow the ideal of physical theory, they must predict and explain everyday activities, using *decontextualized* features. But since the context in which human beings pick out the everyday objects and events whose regularities theory attempts to predict is left out in the decontextualization necessary for theory, what human beings pick out as objects and events need not coincide with those elements over which the theory ranges. Therefore predictions, though often correct, will not be reliable. Indeed, these predictions will work only as long as the elements picked out and related by theory happen to coincide with what the human beings falling under the theory pick out and relate in their everyday activities.

My thesis can be illustrated by the sort of difficulties that confront Lévi-Strauss's structuralist theory of gift exchange. Pierre Bourdieu, in *Towards a Theory of Practice*, argues that Lévi-Strauss's formal, reversible rules for the exchange of gifts – abstracted as they are from everyday gift giving – cannot account for and predict actual exchanges. His point is not that theory leaves out subjective, so-called phenomenological, qualities of gift exchange. That would not be a valid objection. The natural sciences legitimately abstract from subject-relative properties. Bourdieu's point is that Lévi-Strauss's abstraction of the pure objects exchanged leaves out something essential since the exchange would not take place at all if it was perceived as the simple, reversible operation Lévi-Strauss's theory describes:

> The difference and delay which the [theoretical] model obliterates must be brought into the model not, as Lévi-Strauss suggests, out of a "phenomenological" desire to restore the subjective experience of the practice of the exchange, but because . . . the *interval* between gift and counter-gift is what allows a pattern of exchange that is always liable to strike the observer and also the participants as *reversible*, i.e. both forced and interested, to be experienced as irreversible.[18]

But this is still not the whole point, since the scientist need not care how the exchange is perceived, or that it must be misperceived if it is to take place at all. The real problem arises because the meaningful context does more than provide a necessary cover-up of the formal skeleton of the transaction. The tempo of the event actually *determines what counts as a gift*:

> In every society it may be observed that, if it is not to constitute an insult, the counter-gift must be *deferred* and *different*, because the immediate return of an exactly identical object clearly amounts to a refusal.[19]

Predictions based only on formal principles fail in those cases in which what formally counts as a gift in the theory is rejected because it is reciprocated too soon or too late to count as a gift in everyday practice. Bourdieu makes this point (although he does not seem to realize its importance):

> It is all a question of style, which means in this case timing and choice of occasion, for the *same act* – giving, giving in return, offering one's services, paying a visit, etc. – *can have completely different meanings at different times*.[20]

What falls outside the social sciences is not the scientists' ontological interpretation of what counts as social interaction, singled out by Taylor, nor the scientists' ontic skill in picking out a particular type of social event, stressed by the ethnomethodologists, but rather the societies' interpretation of what counts as the sort of event in question. *The meaning of the situation plays an essential role in determining what counts as an event, and it is precisely this contextual meaning that theory must ignore.*

John Searle argues that formal, causal accounts in the social sciences must fail because intentional mental states play a causal role in human behavior and so must be taken account of in any science of human capacities.[21] But since much of human behavior could and does take place as ongoing skillful coping without the participants having any mental states (i.e. beliefs, desires, intentions, etc.) at all,[22] intentional causation does not seem to me to be the right place to look for an essential limit on prediction in the human sciences. What is crucial is that, even when no intentional states are involved, what human beings pick out as specific sorts of objects depends on background skills which are not rule-governed, and so what counts as a certain sort of object is not rule-governed either. Thus prediction will fail whenever an object that according to the theory has the defining features of a given type of object is not counted by those in the culture as belonging to that class. Imprecision in the study of human capacities is inevitable because what counts as an everyday fact depends on a background of meanings and skills which is explicitly excluded by the decontextualization required by theory. Prediction will fail every time the situation is abnormal since in such cases only *ceteris paribus* rules which presuppose the skills of living and acting in a particular culture, not the strict rules required for prediction, determine what counts as the facts.

Predictive failure is a constant possibility in any area where features such as being a gift depend on background practices not integrated into the theory. The problem is not that the social theorists do not have a theory of how they (the theorists) determine what counts as a gift – the fact that they have no such theory is no more damaging than that physicists do not have a theory of how they classify bubble chamber photographs. The problem is that any attempt at a context-free definition of a gift will not coincide with the pragmatic way the social situation defines a gift; that social theorists do not and cannot have a *theory* of how the people they are studying determine what counts as a gift; and, therefore, that predictions concerning gift-giving can at any moment fail. There is no corresponding problem for bubble chambers since bubble chambers do not classify their own bubble-tracks.

The natural way to cope with this problem is to abandon the attempt to find what one might call first-order interpretation-free features and to settle for second-order features such as the participants' judgments as to whether what was just exchanged counted as a gift. Such judgments or, more generally, scores on questionnaires, would be treated as context-free facts or features and then related by theory to explain and predict other objectively determined judgments or test scores. Economics attempts to be such a second-order science. It accepts the current understanding of money, property, etc. and current desires to maximize possessions etc. It then seeks laws relating these socially defined elements. As long as the practices defining the objects and goals of a group of people remain constant, economic laws can in principle be as predictive as those of physics. But if background conditions determine, in a way which in principle falls outside a theory, what counts as the events over which the theory ranges, the theory is at the mercy of changes in these conditions which at any moment can undermine the predictive power of the theory.

It would seem that if one made explicit the theory's boundary conditions one could at least predict when the theory would fail. Thus ecological theories are explicitly understood as applying only as long as the temperature of the earth remains relatively constant. And, indeed, if the conditions which form the background of human sciences such as economics could be made explicit, one could gain a kind of closure by stating the limits within which a given economic theory was meant to be predictive. But in the human sciences the background conditions are not physical facts, nor are they psychological facts such as what agents desire and what they believe is rational.[23] In the human sciences the background conditions are not systems of beliefs and particular intentional states but patterns of behavior which embody an ontology. As Bourdieu says, such patterns of habits go without saying because they come without being said. The problem of the human sciences is that such background conditions change yet the social scientist cannot state in advance what aspects of these patterns have to remain constant for his predictions to continue to be fulfilled. As Taylor has pointed out, objective political science was unable to predict or understand the changed interpretation of society expressed in the hippy movement. In general, since the boundary conditions of the human sciences are not facts but interpretations, social sciences such as economics which build on second order judgments that presuppose human self interpretations are not only incomplete and unstable in comparison with ideal theoretical disciplines such as physics; they are incomplete and unstable even when compared to disciplines with changing background conditions such as ecology.

Philosophers of science differ on what counts as an explanation, but one thing seems clear: if all the factors which the theory takes into account can remain the same, yet the resulting behavior vary, such an account is not a complete explanation of the behavior in question. This leads to the social scientists' search for first order features such as those sought by behaviorism or cognitivism. The social scientist who wants a stable predictive discipline like those in ideal natural science must seek elements which can be recognized and decontextualized without the aid of the participants' everyday practices and discriminations. But if my argument is right there are no such features. All features are contextually determined. Thus, paradoxically, cognitivism's attempt to avoid the limits on the predictive power of second-order social sciences such as economics by grounding all social sciences in a theory of individual capacities, results in a theory whose predications can fail at any moment even where the background conditions remain constant.

Foucault's historical observation that the sciences of man are driven to try to objectify their own conditions of possibility, obscures the real basis of their instability. What drives the human sciences to try unsuccessfully to objectify their background practices is not merely a particular historical structure defining what counts as their object as Foucault holds, but the requirement that any predictive science must bring all the relevant factors into its theory, or at least be able to set out in advance what such factors are. That is why would-be sciences such as structuralism and cognitivism that no longer study human beings as subject/object, or transcendental/empirical doubles are still caught in the necessity and impossibility of giving a theoretical account of the pragmatic conditions which determine in each situation what will count as the sort of features and events whose relations the theory seeks to predict.

Thus finally the human sciences are driven to treat the whole meaningful context as a belief system, where beliefs are thought of as computational states made up of isolable features related by strict rules. Rorty's optimism concerning Cognitive Science must stem from his conviction that some such development is possible. And, indeed, if the background consisted of shared beliefs there would be some hope that it could be captured in a cognitivist data structure. But Heidegger's description of everydayness in *Being and Time* and my description of skills suggest that the pragmatic background cannot be understood as a belief system. In some cases our basic social skills may have been acquired by consciously following rules defined over elements but, like any skill which has reached a level of mastery, our ability to cope with everyday things and situations is no longer caused by and cannot be analyzed into the elements which went into its acquisition. Yet no one has found any other elements or is even trying to find them. Thus recent attempts by AI researchers to analyze everyday, background understanding in terms of elements and rules face the same problems which confront any attempt to produce a theory of a specific human capacity.

If everyday human behavior cannot be predicted by laws or rules relating elements abstracted from that behavior, we can understand why the new structuralist and cognitivist sciences of human capacities have turned out to be as unstable and non-normal as were the old behaviorist sciences of man. All such theories must either refuse to make predictions or else confront repeated exceptions. Given the lack of any of solid success, other approaches inevitably arise in the discipline which offer competing types of systematic accounts. Such competing types of account do not agree on method, evidence, or even on what are the problems. We then have neither normal science nor even revolutionary science, but just the sort of pre-paradigmatic instability characteristic of the social sciences from their inception to the present.

4. The logical possibility of a theory based on other features than those abstracted from the everyday context

It follows from the above considerations that the right vocabulary for what were once called the human sciences would have to be a vocabulary which picked out entirely different features than those abstracted from our everyday activity – features or attributes which would remain invariant through changes in pragmatic situations and across cultural revolutions. Just what such features would be, no one can say. And for this very reason there is no way to prove that such hypothetical features do not exist. All agree that if there were such features and we were able to find them by some sort of luck or divination, then the study of human capacities could, in principle, be closed, predictive, and normal. Entertaining such logical possibilities has little use, however, except to keep someone like Taylor or myself from arguing that, *in principle*, the study of human behavior can never be scientific. But from the fact that there is no in-principle argument against the possibility of a scientific theory of human capacities, one cannot conclude that the instability of the social sciences is an historical accident. Even less can one conclude that current cognitive science may, at last, be on the right track.

Nor do such in-principle arguments give any support to Rorty's claim that what is holding up the development of normal social science is the complexity of their object.[24] Looked at historically the social sciences make a quite different impression than sciences such as biology or meteorology that are struggling with a very complex domain. These latter sciences are making slow but steady progress developing more and more complicated theories taking into account more and more factors, whereas the human sciences first seek to develop a general theory using one sort of feature, e.g. units of behavior, then scrap that attempt and take up some other sort of feature, e.g. the cues sought by cognitivism, without showing any progress at all.

Taylor and Rorty, because each seems to think at times he has proven more than he has or could, fail to appreciate each other's points. Taylor sometimes seems to argue that human self-interpretation plays a causal role in human behavior and therefore we cannot predict across changes in self-interpretation. At other times Taylor makes clear that he has a weaker argument. He admits that in principle one might have a science of human beings under some other description than that of agents, but he argues that to formulate a theory of human activities in other terms that those abstracted from everyday activities would be to deny the reality of human action as we understand it.

Rorty's optimism concerning cognitive science shows that he too thinks his arguments prove more than they can. He sometimes seems to think that he has shown that since there is no in-principle argument against a science of human capacities based on context-free features, any science of human capacities, even the sort arrived at by cognitivists and structuralists, might work. That is, he seems mistakenly to conclude from his correct point that there could in principle be a successful science of human capacities, that no argument could show that the human sciences as currently understood and practiced cannot succeed.

I think the argument I have presented enables us to see why a human science based on features abstracted from everyday life cannot succeed. Either current social sciences must, like economics, build on a background they cannot adequately describe and so whose changes they cannot predict, or else, like behaviorism and cognitivism, human sciences must use features obtained by abstracting from pragmatic contexts to predict events in pragmatic contexts. But since the context determines what counts as a feature this attempt at completeness is bound to fail. This leads cognitivism, especially artificial intelligence, to try to formalize the everyday, common-sense context itself. But this project in turn must fail since background skills cannot be understood in terms of the features they enable us to pick out. (Whether the background skills can be analyzed in terms of some other features we know nothing of is not in question since we have no access to such features and they are not sought by current cognitive science.)

5. Conclusion

To clarify this complex debate I will separate the issues into four distinct questions:

(1) Can there be a science of *man*? Here Foucault is right. There can be no stable science of an entity which as meaning giver is the condition of its own objectification. No science can objectify the skills which make it possible. But this only shows we should abandon the Kantian definition of man.

(2) Can there be a science of human capacities? Here Taylor and Searle have half the answer. All social sciences presuppose human self-interpretations and are only as stable as those interpretations. Moreover, as I have sought to show, the cognitivist attempt to remove this difficulty by seeking context-free features only makes matters worse. If one seeks a stable theory of human

capacities modeled on the ideal of the natural sciences, such theories, which must abstract context-free features from everyday contexts and then predict everyday human activities using rules or laws relating these features, will fail to map onto the context-determined activity they seek to predict.

(3) Can there be a theory of the background practices themselves? Although one cannot prove this is impossible, a phenomenological description of skill acquistion suggests that the features and rules used in acquiring skills play no role in the exercise of expertise, and no one has proposed any other features nor shown why we should expect there to be any.

(4) Can there be a science of human capacities using other features than those abstracted from everyday practice? Here Rorty is right. In principle such a theory is possible. But, one must add that since we have no precedent for such a theory, no reason to believe the abstract features it would require exist, and no way to find them if they did, this abstract philosophical point casts no light on the past, present or future difficulties facing the social sciences.

6. Epilogue

If a theory of human capacities using features abstracted from the everyday world can be expected to fail, what can social scientists do? On a macro level they can go on doing economics and political science accepting their relativity to background practices and so their inevitable instability. But human scientists must also deal somehow with individual human capacities such as language, perception, action, etc. One thing they can do in this area is go on as behaviorists and cognitivists have, trying to abstract features so as to develop closed predictive theories, pushing each new theory to the limit until it fails to predict some phenomenon it is supposed to cover, and then jumping on the next new bandwagon to come along. Some useful data accumulates in this way, although not very much.

Another possibility is to develop a disciplined study of human beings which does not seek to be a *theory* on the model of ideal physical science but still seeks to be a systematic account of everyday activities. One such possibility is the study of the role of prototypes in various areas of human activity. Kuhn has shown that the background practices of the natural sciences are organized by exemplars, whose role can be studied by the historian of science. Clifford Geertz and Foucault organize their interpretations of cultures around specific paradigms such as the cock fights in Bali or European prisons and confessional practices which focus general characteristics of the background practices. It may be a coincidence, but I think it is not, that some linguists are now studying the prototypes which show up in the grammars of all languages. For example, by stuying transitivity in many languages, linguists have been able to show that in a prototypical transitive sentence the subject is an agent, the object is inanimate, and the relation is one of transmission of causal power. On this basis one can predict that a transitive sentence like "John desires a change," would be relatively rare in the world's languages and relatively hard to learn. This turns out to be true.[25] The study of prototypicality effects is also a lively branch of current psychology. Thus research has shown that the study of prototypes in history of science, anthropology, linguistics, and psychology can lead to illuminating results and even to repeatable experiments.

Yet the function of prototypes cannot be captured in a theory. A scientific paradigm, Kuhn has pointed out, is neither rationalized nor rationalizable. An activity is recognized as good science by being seen as similar to an exemplar, without it being similar with respect to any abstractable, identical features. Thus paradigms or exemplars cannot be treated as abstract types; rules cannot replace the typical case. More generally, human beings tend to agree in their judgments of how similar an object or event is to a prototypical case, without being able to rationalize their judgment in terms of context-free features as required by cognitivist theory. Indeed, as Eleanor Rosch has noted, what counts as a relevant feature seems to follow from judgments of similarity rather than to be presuppoed by them.[26]

One could, of course, try to account for judgments of similarity by postulating some sort of highly complex features of prototypes abstracted by the mind in ways unknown to the human beings doing the abstracting. But this is not how cognitive science now proceeds. Rather, an army of investigators have been searching for the sort of abstract features which human beings do on occasion actually use, in an attempt to use such features in a theory which would explain similarity judgments. As one would expect there has been no success.

Meanwhile, prototype studies are producing repeatable, unified accounts of human capacities in perception, knowledge, language, etc. These are not explanatory theories in the strong sense laid down by Plato, Descartes and Kant, which have succeeded so impressively in chemistry and physics. Yet such accounts are as scientific as some areas of biology such as current accounts of evolution. After 2000 years it seems clear we must give credit to Socrates and Plato for the vision of theory which has flourished in the natural sciences, but in the human sciences it might turn out that Euthyphro, who kept trying to give Socrates paradigm cases rather than abstract rules, was a true prophet after all.

NOTES

1 An earlier version of this paper was presented at the *Boston Colloquium for the Philosophy of Science*, October 1982. I wish to thank Mark Okrent and Ned Block for their helpful comments on that occasion.
2 *The Review of Metaphysics*. Sept. 1980, pp. 3–55.

3 Michel Foucault, *The Order of Things*, Vintage, 1973, p. 348.
4 Charles Taylor, "Interpretation and the Sciences of Man" in *Interpretive Social Sciences*, ed. Paul Rabinow and William M. Sullivan, University of California Press, 1979.
5 Mark Okrent, "Hermeneutics, Transcendental Philosophy, and Social Science," *Inquiry*, 1984 (forthcoming).
6 Richard Rorty, *Philosophy and the Mirror of Nature*, Princeton University Press, 1979, p. 352.
7 Foucault uses "human sciences" to denote a special sort of discipline which derives from Kant and which studies what Foucault calls "man". I have been using "human sciences" more generally to refer to any discipline which studies human beings. In this section discussing Foucault, I will use the term "human sciences" in his more restricted sense.
8 Michel Foucault, *The Order of Things*, Vintage, 1979, p. 348.
9 This claim is implicit in the work of Harold Garfinkel, but, as far as I know, has never been set forth as an explicit argument. See Harold Garfinkel, *Ethnomethodology*, Prentice-Hall, 1967, Chapter 8.
10 Op. cit., p. VII.
11 One can, of course, recall the rules one once used and act on them again, but then one's behavior will be halting and clumsy just as it was when one mastered the rules as an advanced beginner. For a more detailed account of the stages of skill acquisition and the implications of this account for cognitive science, cf. *Mind Over Machine*, Hubert and Stuart Dreyfus, Free Press, 1986.
12 Ibid., p. 364.
13 For a discussion of Foucault's hopes and later disappointment in this area see, H. Dreyfus and P. Rabinow, *Michel Foucault: Beyond Structuralism and Hermeneutics*, Chicago University Press, 1982.
14 See *What Computers Can 't Do*, Harper and Row, 1979, for my critique of cognitivism, and *Michel Foucault: Beyond Structuralism & Hermeneutics* for my critique of structuralism.
15 We will have to speak of human *capacities* since the idea of a science of human *behavior* is as much a reflection of one side of the transcendental/empirical double as a science based on mental *representations* is of the other.
16 In asserting that theories range over decontectualized, uninterpreted facts I am not endorsing a positivist account of theories as based on brute data. The basic data are meter reading and computer print outs. These "uninterpreted facts" are cut off from our everyday world of equipment and purposes but they are, of course, recontextualized or theory laden, otherwise they could not fill in variables asking for mass, charge, etc.
17 Martin Heidegger, *Being and Time*, Harper and Row, 1962, p. 194.
18 Pierre Bourdieu, *Outline of a Theory of Practice*, Cambridge University Press, 1977, pp. 5, 6.
19 Ibid., p. 5. My italics.
20 Ibid., p. 6. My italics.
21 John Searle, "Intentionality and its Place in Nature," *Dialectica* (forthcoming).
22 The best argument for this view is contained in Heidegger's, *Being and Time*, Chapter 3. I try to spell out and defend the argument in *Being-in-the-World: A Commentary on Division I of Being and Time*, MIT Press (forthcoming).
23 cf. John Searle's claim in his BBC Reith Lectures, 1984.
24 Private correspondence.
25 Paul J. Hopper and Sandra A. Thompson, "Transitivity in Grammar and Discourse," *Language*, Volume 56, Number 2 (1980).
26 Eleanor Rosch, "Principles of Categorization," in *Cognition and Categorization*, ed. Eleanor Rosch and Barbara B. Lloyd, Erlbaum, 1978, p. 42.

NARRATIVE VERSUS ANALYSIS IN HISTORY

W.H. DRAY

Over the last three decades, historians have often argued vigorously with each other about whether their aim should continue to be the traditional one of elaborating narrative accounts of selected developments or periods of the past; and their disputes have been loud enough to be heard well beyond the confines of their professional conclaves. The *avant garde*, opting for something newer and better than mere "story-telling", which they generally call "analysis", have sometimes gone so far as to question whether those who continue to narrate can be said really to convey significant knowledge of the past; and there has been talk of the demise of narrative as part of a "breakthrough" in historical method of an order comparable to the one which, now hallowed by the name of Ranke, occurred early in the nineteenth century and first encouraged historians to think of their inquiries as constituting a "science". Lawrence Stone, not many years ago, was prepared to say at least that all the really important historical works of the post-war period have been analytical in character, and to praise them for so being.[1] More traditional historians like J.H. Hexter and Geoffrey Elton have responded with what are now often seen, at any rate so far as the relevant literature in English is concerned, as classical statements of the continuing credo of the narrative historian.[2] Not all philosophers who have reflected upon the knowlegde-claims characteristic of particular disciplines have recognized an obligation to take account of the self-conceptions of those who practice them. But philosophers of history who do recognize such an obligation – and their number seems to be increasing – can hardly fail to be interested in this controversy.

Of course, in recent years, analytical philosophers who write about historical knowledge have been pursuing their own inquiries into the idea of historical narrative. Morton White was the first to ask what the function of narrative was in historical inquiry – or, rather, since his major concern was with the logical structure of explanation as it might be sought in different domains, he asked in what way an historical narrative might explain.[3] His answer was that, although other structural features of narrative – for example, its having a central subject or its contents being selected in accordance with appropriate

Margolis, J., Krausz, M. and Burian, R.M., (eds.), Rationality, Relativism and the Human Sciences.
 ISBN 90-247-3271-9.

criteria of significance – would have to be considered in any overall account of it as an intellectual form, so far as it achieved explanation, it would have to do this by offering a *series* of explanations of its constituent events, these all necessarily being of the well-known covering law type since only that type, according to White, was logically sound. In fact, on White's view, the ideal explanatory narrative would assume the form of a causal chain. A.C. Danto considerably improved this causal model by pointing out that particular causal explanations are characteristically given in terms of intrusive conditions.[4] In accounts of changes from antecedent to subsequent states of a development being narrated, causes generally come in, as it were, from the outside. They would thus seldom allow the examined development itself to be represented as a chain of causes and effects. W.B. Gallie modefied White's model further by arguing that what allows a would-be explanatory narrative to proceed from point to point is often reference to conditions which, although perhaps necessary for what occurred at each point, and in some relevant sense explanatory, are still not properly described as causes.[5] For example, they might simply be conditions that provided certain agents with opportunities for doing what they did. An historical narrative may thus make a series of happenings understandable in the sense of being followable without puzzlement, even of seldom – and perhaps never – trying to show what happened was more than contingent.

There were ideas in both Danto's and Gallie's responses to White, however, which offered a challenge, not only to the latter's unrealistic model of a causal chain, but also to his more fundamental, and, I think, more persuasive notion that the understanding afforded by a narrative is just a function of the individual explanations which it contains, however varied these may turn out to be in logical type. Gallie made reference, although perhaps too vaguely, to the sense a narrative commonly conveys of individual episodes contibuting to an as yet unknown but promised overall significance. Danto, bringing squarely into his analysis the crucial consideration that historical accounts are written with hindsight, drew attention to some of the ways in which the characteristic language of historical narrative presupposes – and perhaps, ideally, presupposes at every point – knowledge not only of what preceded but of what followed. It is Louis Mink, however, who has developed furthest the idea that, in consequence, an historical narrative normally has a synoptic aim – a feature which it shares with the classical novel.[6] The most important thing a narrative does to make its subject matter intelligible, Mink maintained, is gradually to bring into view a larger whole, this by means of various literary techniques, such as the use of temporally overlapping descriptions – another way of making Danto's point. Mink went on to contrast the synoptic understanding which narratives seek with the sorts typically aimed at in philosophy and in science. Philosophers typically ask: What is the true nature of this thing? Scientists typically ask: Under what generalization or general theory does this thing fall? Historians typically ask: How does this thing fit into its particular context, past, present and future? Distinctively philosophical understanding could thus be called *categoreal*, distinctively scientific *theoretical*, distinctively historical *configurational*. And configurational understanding, Mink maintained, is characteristically achieved by narrative.

I rehearse all this for two reasons. First, I don't think that what the philosophers have been doing has always been adequately understood by historians – even by those who, like Hexter, have taken the trouble to read them. Their conclusions have often been seen – especially when they address such problems as the nature of historical explanation or historical narrative – as too aridly formalistic to make useful contact with historians' attempts to render past events and conditions concretely intelligible. The way Hexter has sometimes phrased this complaint is that philosophers have failed to take account of the *rhetoric* as well as the *logic* of historical work. Yet much of what he means by rhetoric appears at least to overlap with that conceptual logic of historical inquiry to which White, Danto, Gallie and Mink, in their several ways, contribute. For example, what Hexter says about the need for "processive" explanations in historical accounts bears a close resemblance to what Gallie had in mind when he claimed that following a narrative often requires, not reference to formally sufficient conditions, but to what the interests and perplexities of a likely reader make relevant at a particular point, these often being, at most, necessary conditions.

More troublesome, perhaps, is Hexter's objection that the sort of analysis of the idea of a causal chain offered by philosophers like White throws no light whatever on a question that would be of the first importance to any historian who was lucky enough to come across such a chain, namely where an account of it should begin.[7] As Hexter rightly points out, nothing in White's formal logic of cause and effect would explain why such a chain should not be pursued backwards indefinitely. To my knowledge, no analytical philosopher of history has in fact seriously asked whether there can be good reasons, even rough ones, traceable to the very nature of the causal concept, for regarding one episode of a chainlike explanation as a more appropriate point of origin than another; and that is certainly unfortunate – the more so since how far back particular causal explanations should be pushed is something that historians argue about a good deal. A recent example is the lively debate still going on, in which both Stone and Hexter have participated, over whether the causes of the English Revolution or Civil War were long or short term.[8] But the question itself, if, in Hexter's terminology, it is one belonging to history's rhetoric, is also, in the terminology of analytical philosophers, clearly one belonging to its conceptual logic.

My second reason for having given a brief resume of what analytical philosophers have been saying about historical narrative is that I shall want to draw upon it from time to time. If I don't set out systematically to criticize or develop it, that is because what I want especially to reflect upon in this paper are some of the things I have found, not philosophers but historians saying. Since Stone and Hexter are both on record on the question of the relative merits of narrative and analysis, I shall sometimes take pronouncements of theirs into account, and I shall draw illustrative material from a field to which both have made distinguished contributions, early seventeenth century English history. What I propose to respond to more directly, however, are certain elements of a rather comprehensive statement by a French historian, François Furet, of the reasons why narrative has come to rest under such a cloud in recent historiography. Furet, of course, works within the *Annales* tradition, where hostility to narrative history has long been the mark, not of an articulate minority as in English-speaking countries, but of the professional establishment. In 1975, he published, in English translation, an article tendentiously entitled "From Narrative History to History as a Problem."[9] In that article, most of the concerns about narrative which I have found scattered through the writings of other historians are forthrightly stated. Furet's complaints include the following: that narrative history achieves only the illusion of explanation, its logic really being "post hoc ergo propter hoc"; that it assigns undue significance to events, and especially to events conceived as unique; that it unduly limits historical accounts to variations upon what the records of the past actually say, and hence to the judgments and recollections of the participants; that it is questionably and incurably teleological; and that it too passively allows the past to dictate the questions historians ask of it. What Furet has to say often seems addressed as much to the way he thinks narrative historians have traditionally performed as to what he sees following from the very idea of historical narrative. Taken either way, however, his contentions appear to me a useful point of departure for any attempt to comprehend, and perhaps to resolve, the narrative/analysis dispute.

I

Let me turn, then, to Furet's first contention: that the logic of historical narrative is *post hoc ergo propter hoc* (p. 110). This may seem a rather strange thing to say; but Furet is not alone in saying it. For example, in contrasting the account of the coming of the English Civil War offered by C.V. Wedgwood with what he calls "the more ambitious constructs of the analytical historians," Stone also refers to her "straightforward *post hoc, propter hoc* narrative."[10] The *literal* interpretation of this would presumably be that although what narrative historians typically write is non-explanatory chronicle, they are commonly read as if offering implicit explanation. There is surely little excuse for reading a chronicle of events in this way, however, unless the historian supplies cues for so doing. Unlike some philosophers, I think that chronicles may indeed be non-explanatory (although still exhibiting some other features of genuine narrative, such as having a central subject and, if only through selection, registering judgments of importance). But narrative histories need not be, and never are in fact for long, mere chronicles. They characteristically offer what might be called "running explanations" of the events they narrate – including the sorts envisaged by White, Danto and Gallie.

It is nevertheless striking how often historians themselves appear ready to reduce the idea of a historical narrative to that of a mere chronicle. Thus Christopher Hill expresses regret that Wedgwood's histories only tell us *what* happened, when what the historian should be doing is *explaining* it – as if narrative, by its very nature, excluded explanation.[11] J.P. Kenyon, in commenting on the tremendous change in Tudor and Stuart historiography from the days of the great nineteenth century narrative historian, S.R. Gardiner, observes that one could hardly expect to find in historical accounts of the latter's day the explanation naturally demanded of contemporary analytical studies. It was necessary first, he says, to establish "what actually happened."[12] Narrative historians often appear to accept rather meekly such characterizations of their work. Thus Wedgwood tells of having made a deliberate choice of writing "what" and "how" history rather than "why".[13] Yet the notion that narrative is, by its very nature, entirely non-explanatory hardly bears examination. Certainly Wedgwood's *The King's Peace* and *The King's War* are studded with what would normally be called explanations of actions, events and circumstances. The "propter hoc" complaints can thus be taken seriously only as the suspicion that, given the exigencies of narration, a narrative historian's explanations will always be deformed or inadequate.

So let us ask whether there is some kind of incompatibility, or at any rate, serious tension, between the ideas of narrating and of adequately explaining a sequence of events. The concern is presumably that narrative, being necessarily linear, is explanatorily thin; analysis, relieved of the obligation to tell a story, can be as thick as you please. To put it another way, in tracing linear sequences, one must ignore the background; to the extent that one allows oneself to be diverted into characterizing the background the narrative loses its thread. But adequate explanation of any transition in a sequential narrative is impossible, it will be held, if the background is thus ignored. So one is forced either to narrate or to explain: one cannot do both. Maurice Mandelbaum put this difficulty in a usefully specific form when he argued that the commonsense notion of a cause normally employed in narrative history is itself enough to ensure explanatory incompleteness.[14] That notion always requires the highlighting of one or a few relevant antecedent conditions belonging to possibly enormous sets, jointly more or less sufficient for what occurred, the unmentioned ones, however, being as necessary for it as the ones mentioned. Mandelbaum thinks causal narration proceeding in this way so certain to mislead that he would repudiate the cause/condition distinction altogether, and historical narration with it.

Two sorts of reponse to this general line of argument suggest themselves. First, it isn't true, without qualification, that narrative historians must ignore the background that falls outside the series of happenings they wish to trace. Narratives are not elaborated in the void; they *presuppose* general knowledge of the background conditions which would fill out the explanations they may offer at various points; and if the adumbration of this background is to be called analysis, they presuppose analysis. In fact, analytical inquiries are sometimes quite deliberately presented as providing, or perhaps as correcting, background knowledge required for associated narratives – as seems to be true, for example, of Hexter's investigation of the political and religious significance of the labels "presbyterian" and "independent" in English Civil War historiography.[15] Sometimes whole analytical works are offered as background for revising narratives: Stone's *Crisis of the Aristocracy*[16] has certainly been read this way, and some of the author's own remarks encourage such a reading (pp. 5, 13). Nor are narrative historians without techniques for calling attention to necessary background knowledge, whether by way of prolegomenon, as in the famous third chapter of Macaulay's *History of England*, or in a more allusive and piecemeal manner. All of which concedes, of course, that sophisticated narrative is not independent of analysis. But it doesn't show that an historical work that places the main emphasis on tracing a linear sequence is to that extent defective or scarcely explanatory. It is so only to the extent that it requires background knowledge that we don't have.

This point needs reinforcement with the reflection that analysis may equally, from time to time, require, and perhaps also presuppose, narrative. First, as has already been suggested, pre-existing narrative histories – or the need to revise them – are what often give particular analytical works their point. But beyond that, analytical accounts whose subject matter spans any considerably period of time often find it difficult to get by without narrating at all. If there is any "lesson" that most historians would agree that history teaches, it is that nothing in human affairs stays as it is for very long. Analytical works – Christopher Hill's *Economic Problems of the Church* is an example[17] – are sometimes accused of forgetting this, tending to treat as homogeneous for purposes of analysis what cannot be safely so treated because of the passage of time. Analysis doesn't have to ignore change, of course; some sorts of change can be shown conveniently on a graph, while others can be indicated by successive cross-sectional soundings. But sensitivity to change generates a propensity to narrate – to trace out the change, to show in detail what it is.

In fact, it is as notable a feature of good analytical historical works that from time to time they pause to narrate, as it is that narrative histories pause to analyze. This is eminently true of those that are overtly explanatory, like Stone's *Causes of the English Revolution*. As Stone himself remarks, an analysis of the Revolution's causes into explanatory factors not related chronologically fails to close the gap between generally predisposing conditions and the actual event: narrative eventually proves necessary if only to show how a society in some danger of catastrophe went over the edge (pp. 135 ff). But the accounts given by Stone of predisposing conditions like economic growth or the rise of a parliamentary opposition themselves tend to offer at least the materials of narrative, and constantly threaten to break out into narrative. Even analytical works more deliberately oriented towards structural description, like *The Crisis of the Aristocracy*, seldom renounce narrative altogether. In the latter, for example, Stone frequently offers mini-narratives like this summary account of the causes of the Essex rebellion: "It was lust for power and responsibility which had lured these men to Court in the first place, thwarted ambition which had driven them to dissipation, dissipation which had caused the mounting burden of debts, and the necessity to reduce the burden which provided the added incentive to an attempt to seize office by force" (p. 221). As Morton White would doubtless be cheered to note, this mini-narrative even traces out something remarkably like a causal chain.

Stone is nevertheless surely right to insist that clear answers to questions like "What were the causes of the English Revolution?" are not easy to find in the writings of classical narrative historians like Clarendon and Gardiner[18] – in spite of the running explanations these contain. He makes a related point when he associates the return to detailed narration in work currently being done on Stuart England with a disinclination on the part of historians any longer to ask "the big why question" (*PP*, p. 81). What he has in mind is a contrast with what was more characteristic of the three preceding decades, when leading historians sought interpretations of the Revolution informed by marxism or other social science theories. But should we really say that what has changed here is the *question*? Isn't it rather the answer that many historians are now giving to it? Contemporary revisionists still surely ask the big why question, but they offer what Stone would regard as little answers – citing, for example, the character of Charles I, or the timing of the Irish Rebellion. What certainly must be recognized is that answers to questions like "What were the causes of the English Revolution?", whether big or little, are not normally set forth *in the course of* a narrative, even by narrative historians. Answering such a question requires an act of retrospective abstraction.[19] What provides the materials for supporting one answer than another, however, may still be an explanatory narrative.

II

In his second criticism of narrative history, Furet points out, as if it were obviously a weakness, that such history typically concerns itself with events regarded as unique (p. 107). What narrative historians write about, he says, is *the* battle of Waterloo or *the* death of Stalin. The implied argument is the familiar one that although all historical events may indeed *be* unique, they do not nead to be *considered* in their uniqueness – the price of so considering them being their remaining inexplicable in any way that could plausibly be called scientific. In rather figurative language for someone calling for a more scientific form of historiography, Furet therefore exhorts historians to cease thinking of historical fact as "the explosion of an important event, which ruptures the silence of time," conceiving it rather as "a chosen and constructed phenomenon whose regularity facilitates its identification and examination . . ." (p. 111). He applies this principle to the demographic history in which he is himself especially interested in the following way. For the demographic historian, he says, the birth of Napoleon "has exactly the same importance as that of any of his future soldiers" (p. 114). The narrative historian can hardly complain that Furet leaves it obscure what his preferred type of history would leave out.

It is a strength of Furet's statement, I think, that it acknowledges as the important question, not whether historical events, unlike, say, events in nature, *are* unique, but rather whether historians may legitimately be interested in them in ways other than as members of designated classes. The battle of Waterloo may be of interest as a battle, a class of historical phenomena about which it is possibly worth theorizing. But we could see it also as an episode in the political history of France or England, with respect to which most other battles would be irrelevant – that is, in Furet's language, we could ascribe significance to it "according to its position on . . . the axis of time" (p. 108). In terms of Mink's epistemological trichotomy, Furet's rejection of this possibility really amounts to a rejection of configurational understanding in favour of the theoretical, or perhaps even the categoreal kind. It is important to see how wide a swath this cuts. For it would be arbitrary to rule out interest in events as unique only to the extent that thet entered into *serial* configurations. Furet's position would thus have to be taken as ruling out cross-sectional configurational studies as well, and, in consequence, much of what is done under the name of analysis: certainly Hill's *Century of Revolution* (excluding its brief narrative introductions), probably Stone's *Crisis of the Aristocracy*, and perhaps even some of that "total" history practiced especially by French historians which attempts to relate in immense detail aspects of the life of a particular village, country or region. I see no reason for any such self-denying ordinance on the historian's part – unless Furet can show that configurational understanding is not really understanding at all.

It might perhaps be questioned to what extent narrative construction does, in fact, require its constituent episodes to be regarded as unique. Certainly – to cite a consideration often underlined by positivist philosophers of history – in using descriptive language at all, the historian necessarily applies general concepts. What these concepts are applied to, however, is not seen as significant merely because it falls under them. A point emphasized by Mink is of interest in this connection. The descriptions under which narrative historians bring events in order to show their configurational significance, he points out, are characteristically of a kind that would not easily find a place in theoretical inquiries. Descriptions like "the birth of the first French Emperor," for example, or "the last time Charles was to see London before returning to it as a prisoner," which derive from the historian's grasp of the way the indicated events enter into narratives, are, to say the least, theory-resistant – as Furet clearly recognized.

The question of how much narrative history concerns itself with the unique arises again at the level of the configurations. Mink has claimed that historical configurations are themselves unique – and not contingently but necessarily.[20] This claim is related to a still earlier contention of his which, more than anything else, I think, helped to liberate analytical philosophers of history from the trench warfare into which their arguments over the covering law model of explanation had settled by the mid-sixties: namely, that the conclusions of historical inquiries, consisting as they do of the entire system of relations which those inquiries reveal, are not detachable from the works that body them forth.[21] Strictly speaking, Mink held, they can neither be summarized nor categorized. I am happy enough with the idea that between any two historical configurations which historians might want to compare there will almost always be found significant differences. However, to stress the uniequeness of such configurations to the point of denying the possibility of overall categorization seems to me questionable. This amounts to a denial that, at the level at which a narrative is grasped as a whole, it is ever proper for a historian to try to impose categoreal understanding upon configurational, or theoretical upon that. One might perhaps have qualms about attempting this at the level of a Toynbee concerned with the pattern of whole civilizations, but surely not – at any rate, in principle – at the level of a marxist interpreting the English Revolution as a bourgeois one, or of a whig regarding the events leading up to it as a response to tyranny in the name of constitutionalism. In fact, it is a common historical practice to bring whole concatenations of events, actions and circumstances under what W.H. Walsh called "colligatory concepts".[22] Even to regard as a "rise of the gentry" what was happening to that stratum of Tudor and Stuart society whose representatives sat in the House of Commons would involve conceptualizing a configuration. It would also involve taking sides – if belatedly – in one of the most famous substantive controversies of modern English historiography.

I have been concentrating upon Furet's reaction to the fact that narrative historians consider it their task to understand events in their possibly unique relations to other events and conditions rather than as falling into classes of theoretical interest. Some of his fire, however, seems to be directed against the idea of historians being much concerned with events at all – a wish to narrate being seen as encouraging such a concern. That narrative history has been largely about events, including, of course, human actions, seems true enough. The conclusion we seem invited to draw is that it has therefore concentrated upon the relatively trivial. In thus taking aim at *l'histoire événementielle*, Furet gives voice to a dominant strain in the thought of the *Annales* historical school, but I think he also speaks for many others who see analytical history as the kind more likely to yield significant knowledge. Not for Stone, however, despite his acknowledgement of some inspiration from *Annales*.[23] It is true that Stone has excoriated the current crop of English Civil War historians – narrativists all – as "antiquarian empiricists", mere "chroniclers of the petty event" (*PP*, p. 93). But I take it that his emphasis was on "petty". In any case, like many other analytical historians, he often finds it necessary, even when not narrating, to talk at length about events. And when he does so, interestingly enough, he frequently talks of them in a language appropriate to narrative (or, as Hexter might say, he employs a "rhetoric" of narrative). For example, in *The Crisis of the Aristocracy*, he calls the period 1580–1620 "the real watershed between medieval and modern England" (p. 12); in *The Causes of the English Revolution*, he points out the "milestones" of the developing constitutional crisis from 1603 (p. 94), and refers to the Revolution itself as "a critical episode in modern English history" (p. xi). Terms like "watershed", "milestone", and "episode" presuppose a serial, narrative understanding of the subject matter to which they are applied. In Furet's terminology, they draw attention to what "ruptures the silence of time."

But how important is it really for analytical historians to take a tough line about history's traditional concern with events? In this connection, it is perhaps worth noting at least the extreme elasticity of the notion of an event. Cromwell's winning the battle of Naseby is obviously an event in any historian's vocabulary. But so, according to Hill, is the Industrial Revolution, which he calls "the most important event of modern history."[24] Furet, at one point, observes that the whole of history can be seen as an event (p. 114). Unless some arbitrary spatio-temporal limitation is placed upon the application of this concept, to say that a narrative approach commits the historian to a study of events therefore need not be very confining. In a pinch, we might even so describe a work like Stone's *Family, Sex and Marriage in England*,[25] which traces the development of the English family over a period of three hundred years from an "open lineage" stage through a "restricted patriarchal nuclear" stage to a "closed domesticated nuclear" one. This work is overtly analytical; but, in Hill's sense at least, it too concerns itself with events. It might also be said to owe something, if only at the most general level, to an embryonic narrative framework. At that level it at any rate exemplifies Stone's own definition of narrative as "the organization of material in a chronologically sequential order, and the focusing of the content into a single coherent story, albeit with subplots"(*PP*, p. 74).

III

Furet's third criticism of narrative history is expressed as follows: "Historic narrative must follow a division of time which is dictated by the crude premisses of experience: basically it records the recollections of individuals and communities. It keeps alive what they have chosen of their past . . . without taking apart or reconstructing the objects within this past" (p. 107). There appear to be echoes of this view in comments sometimes made by other historians who have reservations about narrative history. Stone, for example, in a study of the English Reformation, remarks: "What really happened to the English in the 1530's – indeed, what really happens to any people at any time – cannot be discovered merely by examining the correspondence of the leading minister" (*PP*, p. 113). To some extent, of course, this can be read as a caution against concentrating too much upon elites – according to Furet, another failing of narrative history (pp. 107–8). But the alternative proposed, namely that the real truth be dug out of local and legal records, suggests his present point as well.

As Furet bluntly phrases it, part of the complaint seems to be that narrative historians follow what R.G. Collingwood called a method of "scissors and paste."[26] They are pictured as simply repeating, or refusing to repeat, what alleged eyewitnesses recorded for posterity. Now if narrative history really were circumscribed in method and interest in this way, there would indeed be good reason to call for methodological revision. But in what sense could it plausibly be said that the task of narration confines the historian to repeating unreconstructed "recollections"? It is doubtless true that much traditional history has depended heavily on what are sometimes called literary sources, and, if only as a starting point, on contemporary accounts of what happened. It might perhaps be argued, too, that without such contemporary accounts, a full-blooded narrative history of many stretches of the past would be difficult, if not impossible to achieve. Hence the thinness of prehistory, based only on what can be eked out of artifacts. But it is surely a great exaggeration to say that, basically, narrative history "records the recollections of individuals and communities'' of the past. No reputable narrative historian uses a mere method of ''scissors and paste''. As Collingwood himself constantly reiterated, good history is based, not on testimony but on evidence – which may well include testimony which is made to reveal what it doesn't itself say.

Any implication that a narrative historian would be bound even by the original agent's views of what was important also needs to be resisted. The retrospectivity of narrative construction will ensure that the historian's own judgments about what was *causally* important in his subject matter will sometimes differ from those of the original agents, if only because he is able to draw upon expertise not available to them. And his judgments about what was *intrinsically* important in it – what in it really mattered most, humanly speaking – can also be expected to diverge in some respects from those of the men and women of the time. For example, a relative unconcern with economic matters on the latter's part need not bind the historian to disvalue them in the same way. The same could be said about the *concepts* in terms of which the original agents made their lives intelligible to themselves. These will certainly have to guide the historian in his attempts to understand why various individuals and groups did what they did; but there is no case for their exclusive employment in his own judgments about what actually happened.

All this has a bearing upon the more general question whether the narrative historian's concepts are, as Furet says, ''constructed'' or only ''received''. In fact, quite familiar concepts of narrative history, like ''Puritan Revolution'' – apparently an invention of S.R. Gardiner in the nineteenth century – are just as much historians' constructs as is Furet's favoured concept of the birth rate – and just as remote from the consciousness of the agents to whose activities and experiences the historian applies them. Seventeenth century Englishmen didn't know at the time that they were making a Puritan Revolution (if they were), any more than their eleventh century ancestors knew that they were establishing the feudal system. In fact, there are reasons – or so some historians argue – for questioning whether the men who made the so-called Puritan Revolution even had the concept of *revolution* as we know it. They certainly had the word; but when they talked of revolution they seem to have meant something entirely different from our notion of it: something essentially backward-looking.[27]

Of course, what I have been saying accepts the Mink-Danto thesis that historical narratives are characteristically retrospective. This is something that distinguished narrative historians have at times appeared to question. Wedgwood, for example, immediately associates her desired status as essentially a narrative historian with the claim that she tries only to recreate the past as it appeared to the agents themselves.[28] Conrad Russell, in what Hexter has called his ''old-fashioned march'' through the parliaments of the 1620's,[29] almost makes it a point of honour to avoid talking at any stage as if he, unlike the agents, knew what was coming next. Herbert Butterfield appears close to adopting the position Furet fathers on narrative history generally when he describes its task as a kind of "resurrection".[30] It is true that he also describes it as "an attempt by literary means to show the deployment of the actual events, perhaps as an ideally situated contemporary might have seen them," thus at least renouncing the aim of representing them as contemporaries did in fact see them. But the envisaged task is still the recovery of the past from its own standpoint.

The theory of such historians is one thing, however; their actual practice is another. Even Wedgwood doesn't succeed in writing as if she didn't know what no contemporary, even an ideally situated one, could have known. The prescient description of Charles leaving his capital noted above was actually a paraphrase of one of her chapter endings.[31] Nor do Russell or Butterfield entirely practice what they preach. In fact, good narrative history characteristically exhibits a tension between what might be called the standpoint of the participant and the standpoint of the hindsighted observer; it entails a constant dialogue between empathy and retrospection. Perhaps retrospective knowledge could, in theory, be excluded from what is actually *said* in a history; that is, a Gallie-type narrative simply showing how the original agents responded to their problems as they arose may, in principle, be possible. It may even, at times, be salutary for historians to urge each other to move back towards something like this for a while – for example, after a period of interpretation in which a given segment of the past has been studied too exclusively for its significance for some later state of affairs. It is apparently with a view to correcting an undue emphasis of this sort (or what is perceived as such) in much work done in the English Civil War, that advocates of writing history "forwards" rather than "backwards", like Elton and Russell, have attacked their predecessors for representing the Tudor and Stuart period as a "high road to civil war."[32] But revisionists commonly exaggerate to make their point; and the idea that history should be, or even could be, written entirely "forwards" is surely, at most, a sometimes useful exaggeration.

It is worth emphasizing that the indispensable, if not in itself sufficient, empathetic aspect of historical narrative is not a feature only of that kind of historical inquiry. The empathetic aspect doubtless shades off to zero in simple reports of changing birth rates and the like; but it is by no means absent from a good deal of what would generally be called historical analysis. An interesting case in point is an account offered in Stone's *Crisis of the Aristocracy* of a seventeenth century landholding arrangement called a beneficial lease, which Hexter has praised as a model of historical reconstruction. Stone's problem was why this, the most common form of agreement for letting land in Stuart England, was one which, from the tenant's standpoint at any rate, *we* should probably ragard as economically irrational. Rather than setting up a schedule of tolerable annual rents after an initial payment of something close to what the prospective tenant could afford, the lease required a huge initial outlay that often put him in a position of great hardship. Stone's explanation of this peculiarity evokes a whole view of the world from the standpoint of seventeenth century English peasants, the most obvious characteristic of whose life was uncertainty. Such men, he says, "positively preferred a few years of misery and hardship while the fine was being paid, followed by a long period at a low rent . . ." (p. 150). The explanation is of a customary action, not a particular one; it presumably belongs to what Elton has called "the thick molasses of social history";[33] but its logic is of the empathetic type. Stone himself has said elsewhere – and apparently with regret – that the sorts of questions that seem increasingly to preoccupy historians nowadays, and which demand knowledge of "what was going on inside people's heads in the past, and what it was like to live in the past," will "inevitably lead back to the use of narrative" (*PP*, p. 85). This may ultimately be so; but as his own work shows, these questions can also guide good analysis.

IV

Furet's fourth criticism is of special interest, not only in itself, but because it brings into view some further features of narrative history, and touches on still another issue which historians often debate among themselves. What gives "meaning" to all narrative, Furet declares, is "temporal finality". In this kind of history, "only the 'final cause' . . . makes possible an understanding of the events of which it is made up" (pp. 108–9). In other words, narrative history, according to him, is incurably *teleological*. Somewhat to my surprise, I have several times found historians of the seventeenth century applying the term "teleological" to each other's interpretations as an expression of censure. Paul Cristianson, for example, makes its allegedly teleological character one of his main complaints against the whig interpretation of the pre-Civil War period in England[34] – an interpretation that contemporary revisionists are especially concerned to show mistaken. Critics like Christianson, however, seem to assume that a narrative can be, and ought to be, non-teleological.

At least part of what lies behind both Christianson's and Furet's concern shows itself, I think, in Furet's further remark that the inclination to write narrative history has declined with a decline in the belief in *progress* (p. 110). He apparently sees more than an accidental connection between these two ideas. Is any general difficulty for narrative history to be derived from such an association of ideas? I suppose it is possible that most narrative historians have in fact believed in historical progress. If so, that is something for which it may be worth finding an explanation – perhaps a psychological, perhaps an historical one. But it is hard to see that writing narrative in any way *commits* a historian to such a belief. An account which traces the decline of a cherished value or which probes the remote origins of a disaster is surely just as viable as one that traces a rise or tries to explain what is regarded as a step forward. A linear historical account, in other words, could display regress as well as progress.

But there is an even more important misunderstanding of the nature of historical narrative implied by Furet's linking it in some essential way to a belief in progress – one which was expressed in an influential form by John Dewey when he maintained that every historical account is concerned with a movement towards some selected outcome.[35] Little reflection is required to see that there are at least three, and perhaps four, different overall patterns available to narrative historians. One is Furet's progressive pattern. This is what we should expect, for example, from any political and constitutional history of the sixteenth and early seventeenth century in England that took as its title some variation on Wallace Notestein's well-known phrase, "The Winning of the Initiative by the House of Commons." What might be called the point of reference for such a history would clearly be the happy outcome: this is what would provide the narrative with its criterion of selection and its standard for judging the relative causal importance of whatever found a place in it. Something very different would be expected of a work entitled, say, "The Aftermath of Buckingham's Ascendancy." Here what would give "meaning" to the narrative (to use Furet's own term) would be the point of origin; and what Furet calls teleology would be turned upside down. Still another type of narrative history would have its meaning or point of reference in its middle – as would be suggested, for example, by a title like "The Rise and Fall of the Leveller Movement." What is said about origins or dissolution, about waxing or waning, in such a work would be relative to the "flourishing" of what is taken as the chief object of interest. Still other narrative histories may have no identifiable points of reference at all, simply setting out to recount what was memorable in a temporal slice of a chosen subject matter. The aim of a work entitled "The Reign of James I" would hardly be to show either the consequences of the King's accession or what culminated in his death. Thus the claim that narrative history is, by its very nature, teleological in the sense of being outcome-oriented, appears to have little to be said for it.

Of course, even an outcome-oriented history need not be judged "teleological" in any sense which implies the operation of "final causes" in a metaphysically mysterious sense: a sense which might tempt employment of the metaphor of the future "drawing" events towards it rather that the past "pushing" them on. Still less need there be any implication that events are *inevitably* drawn to some future outcome. An outcome-oriented history need not, in fact, succumb to the thesis of inevitability at all – although the epithet "teleological", when used by historians as a term of reproach, nearly always seems to be taken as implying it. Nor, of course, need a history written on the outcome model imply that nothing of importance was happening other than what led to the designated outcome. Many things omitted from such an account may be acknowledged to be important in some general historical sense. It is just that they are not judged to be imported for the particular problem treated.

Lack of clarity on such points seems to me sometimes to infect disputes between historians about whether a certain account has fallen into a "whig interpretation" of the events it traces. Following Butterfield,[36] I take such an interpretation to mean, first, one that views past events as if their importance lay solely in their tendency to produce some result considered valuable in the historian's own day. Taken in this sense, "whiggism" can surely be perfectly innocent provided the historian's *question* is how much a past period had the seeds of a specified future development in it (rather than, for example, what that period was like). And this despite the endemic dangers of such an approach, like failing to see what was there because of not expecting to find it, or crudely reading back what was true only of later times.

The other main characteristic of what historians often call "whiggism" – ascribing inevitability – is, I think, a permanently seductive feature of the retrospective standpoint. Because things turned out a certain way, it is all too easy to assume that that was how they *had* to turn out – or, to put it in a fashion more likely to be called teleological, the way they were *going* to turn out. I fear that philosophical analyses of the idea of narrative are sometimes presented in a way that could encourage, and even seem to legitimize, what I am calling a seductive illusion. I don't think, for example, that Mink would accept an inevitability theory of history, and still less a theory of specifically teleological inevitability. Yet in the course of the argument in which he very properly advances beyond Gallie's analysis of historical understanding – the idea of following events in a frame of mind very like that of the original agents who, since they had to act, necessarily saw their future as open – he may easily be taken to mean that, once things have happened, the course events took must be viewed as having *had* to be that way. In reconstructing the past, he says, historians trace backwards developments which were seen as contingent going forwards; but "there are no contingencies going backwards."[37] In a summary statement which, I think can hardly fail to give comfort to teleological inevitabilists, he adds: "thus what may be contingent in the occurrence of events is not in their narration." Events which are judged by the historian to have occurred contingently, however, ought surely to be represented in narratives as contingent, no matter how much the techniques of sophisticated narration superimpose retrospectivity and synthesis. The point of substance behind what Mink has to say is surely no more than that if an earlier event was necessary for a later one to occur, then that later event was sufficient for the earlier one's having occurred. Or perhaps better: if reasoning forwards, an event or condition is seen as contributing to the intelligibility of an explanatory narrative by being necessary for a selected end-result, then reasoning backwards from the end-result, that very same condition can be asserted certainly to have occurred. But this is only a claim about the nature of conceivable arguments from evidence, not one about the way a historical narrative would necessarily represent the course of events.

Of course, in a more general, and surely quite innocuous sense, historical narratives are typically teleological in that, being concerned with human actions, they are concerned with the way intentions and purposes enter into the explanation of the past. That is not to say, of course, that such accounts deal only, or even chiefly, with what past agents brought about intentionally: it is not to recommend a conspiracy theory of history. Historians do, from time to time, have to trace conspiracies; and short of that, they commonly have to reconstruct enterprises or movements which exhibited a remarkable continuity of purpose, even when their membership was not continuous. Accused of seeing too much purposive continuity of this sort in the activities of the parliamentary gentry of the 1610's, 20's and 40's, Hexter's surely satisfactory reply was that this was the overlapping sort found in a rope.[38] Even the understanding of unintended historical consequences, however, may have as prerequisite an empathetic grasp of the intentions of those whose actions composed them.

V

There remains the rather strange criticism implied by the title of the article by Furet to which I have been responding: "From Narrative History to History as a Problem." Furet is quite explicit about what he would regard as an historical problem: it would be something like charting the price level in England in a specified hundred year period or comparing the productivity of slave and free labour in the United States before the American Civil War (p. 122). Apparently piecing together the events historians sometimes refer to as the Puritan Revolution or the Great Rebellion or the English Revolution (and finding out which was the more apt way of characterizing them) would not count as solving a problem at all. There would seem to be echoes in this of Lord Acton's advice to "study problems, not periods" – with its surely false implication that periods cannot be problems. Anyone wishing to claim for history the status of a science in some honorific sense (even if not one that entirely assimilates its concepts and methods to those of the natural sciences), will, of course, want to emphasize the extent to which historians are *active* in inquiry. As Collingwood liked to say, they must put the record "to the question."[39] But the Furet-Acton way of applying this idea is surely quite bizarre. Some questions taken to the record explicitly require narrative answers. Others are found to require them once the record is carefully examined.

The idea is nevertheless well entrenched that the work of narrative historians is divided into two parts: on the one hand, the establishment of the facts, which, at its best, is a matter of highly skilled research; on the other, the construction of narratives out of these facts, which is a matter of literary art. Among anti-narrativist philosophers, this position has recently been restated by Mandelbaum, whose case against narrative history is, finally, the complaint that narration is not an *inquiry*;[40] and by C.B. McCullagh, who sees narration as no more than "a dramatic means of describing historical events.[41] A generation ago, the analytical historian, F.J. Teggart, expressed a similar view when he insisted that "with whatever care the facts are sifted, and with whatever sincerity they are subsequently presented, narrative statement remains art, and as such, not science."[42] Unfortunately for the reputation of narrative history, its defenders often talk as if they thought in terms of the same dichotomy. Thus Elton, at one point, opines that whereas analysis is a work of "ratiocination", narrative is one of "imagination".[43] He hardly compensates for this unflattering division of labour by adding that it is imagination which is in shorter supply.

What all such talk of science versus art is in danger of obscuring, I think, is that the configurations outlined by historical narratives are themselves historical facts to be established. Their elaboration does not simply decorate the historian's real conclusions: they are themselves among the conclusions sought – which, if Mink is right, may not be adequately expressible in any other way. The untenability of sharply separating narrative construction from historical *inquiry* shows itself, among other things, in the *directive* role, already noted, which a received narrative often plays in historical research. The research is undertaken because an existing account seems unsatisfactory in certain ways; its results are the emendation of that account. In Mink's terminology, the narrative functions as a "cognitive instrument". Much of what Hexter has to say about narrative as an essential element of history's "rhetoric" seems aimed at making a similar point. The term "rhetoric", however, suggests a communicative art, a literary skill; and what Hexter sees as falling under it does sometimes seem more concerned with how truth is best conveyed than with what its content is. To take a single example: when a narrative should change scale – one of the "rhetorical" problems Hexter considers[44] – seems to me chiefly a literary problem. By contrast, where a causal account should begin – a problem noted in the introduction to this paper – seems to me rather one of content: one concerning the nature of the process under examination. Hexter maintains that the rhetoric of history, far from being a mere literary icing on the historical cake, is "mixed right into the batter."[45] I find this more persuasive in the second sort of case than in the first.

If the tendency to separate narrative from inquiry survives all such considerations, I suspect this may owe something to an underlying ontological worry that surfaces from time to time in the literature on narrative history. Stories, it is tempting to say (with Charles Beard), are just the historian's personal "selections and arrangements of facts." The real historical past is not composed of stories – "untold stories" waiting to be told by historians. Even as committed a narrativist as Mink bridles at the notion of an "untold story". This, he says, is as nonsensical as the notion of "undiscovered knowledge."[46] Perhaps the word "story" is unfortunate here. But shouldn't we be prepared at least to say that the serial configurations which historical narratives delineate really characterize the historical process? And isn't this all that the more daring form of words really intends to convey? The notion of an as yet undiscovered historical configuration is surely no more problematic than that of an as yet undiscovered scientific law. "There is a story waiting to be told" is thus not a form of words entirely lacking in point and substance. The danger of maintaining the contrary is that of relapsing into some virtually unintelligible notion of the human past as a process any abstraction from which would necessarily be a falsification. Worked through, such a notion would put in jeopardy much more than narrative history.[47]

NOTES

1 *The Past and the Present*, London: Routledge & Kegan Paul, 1981, p. 21. Further references to this source are shown in the text by page numbers prefixed *PP*.
2 See, for example, Hexter's *The History Primer*, New York: Basic Books, 1971 (hereafter referred to in the text as *HP*), especially ch. 6, and Elton's *Political History*, London: Basic Books, 1970.
3 *Foundations of Historical Knowledge*, New York: Harper & Row, 1965, ch. 6.
4 *Analytical Philosophy of History*, Cambridge: Cambridge University Press, 1965, chs. 7, 8, 11.
5 *Philosophy and the Historical Understanding*, New York: Schocken Books, 1964, chs. 1–5.
6 "History and Fiction as Modes of Comprehension," *New Literary History*, I, 1970, pp. 541–558.
7 *The History Primer*, p. 166.
8 See especially Stone's *Causes of the English Revolution, 1529–1642*, London: Routlegde & Kegan Paul, 1972, ch. 3.
9 *Diogenes*, Spring 1975, pp. 106–123.
10 *Social Change and Revolution in England, 1540–1640*, London: Longman Group, 1965, p. xxii.
11 Review of *The King's War, Spectator*, 12 December, 1958.
12 Review of Stone's *Crisis of the Aristocracy, New York Review of Books*, VI, 1966, p. 24.
13 Ved Mehta, *The Fly and the Fly Bottle*, London: Weidenfeld, 1962, p. 159.
14 "A Note on History as Narrative," *History and Theory*, VI, 1967, p. 417.

15 *Reappraisals in History*, London: Longmans, 1961, ch. 7.
16 Abridged edition. London: Oxford University Press, 1967.
17 See G. Elton, *The Practice of History*, London: Fontana, 1969, p. 164.
18 Reply to a review by H.G. Koenigsberger in *The Journal of Modern History*, 46, 1974, p. 107.
19 J.P. Kenyon performs such an act on behalf of Gardiner and Wedgwood when he summarizes their view of the Civil War's causes as "a defect of personality and government on the part of the Stuarts" plus "an upsurge of precocious democracy riding on the back of militant Puritanism," in a review of Stone's *Crisis, New York Review of Books*, VI-2, 1966, p. 23.
20 "The Divergence of History and Sociology in Recent Philosophy of History," in *Logic, Methodology and Philosophy of Science* IV, eds. P. Suppes et al, North Holland Publishing Company, 1973, p. 740.
21 "The Autonomy of Historical Understanding," *History and Theory*, V, 1965, pp. 38ff.
22 *Introduction to Philosophy of History*, London: Hutchinson, 1958, pp. 59ff.
23 Reply to Koenigsberger, *op cit.*, p. 106.
24 Discussion with Lawrence Stone and Peter Burke, *The Listener*, 4, October 1973, p. 448.
25 London: Weidenfeld & Nicolson, 1977.
26 *The Idea of History*, Oxford: Clarendon Press, 1946, p. 257.
27 See, for example, V.F. Snow, "The Concept of Revolution in Seventeenth Century England," *Historical Journal*, V, 1962, pp. 167–190.
28 *The King's Peace*, New York: Macmillan, 1956, pp. 15–16.
29 "The Not-So-New-Men," *New York Review of Books*, 18 December, 1980, p. 58.
30 "Narrative History and the Spade-Work Behind It," *History*, LIII, 1968, pp. 165–166.
31 *The King's War*, London: Fontana, 1966, p. 58.
32 See, for example, G. Elton, "A High Road to Civil War?", in *From the Renaissance to the Counter-Reformation*, ed. C.M. Carter, London: Random House, 1965, pp. 325–347.
33 *Political History*, p. 156.
34 "The Causes of the English Revolution: A Reappraisal," *Journal of British Studies*, 15, 1976, pp. 41, 52.
35 "Historical Judgments," in *The Philosophy of History in Our Time*, ed. H. Meyerhoff, New York: Anchor Books, 1959, p. 167.
36 *The Whig Interpretation of History*, London: Bell, 1951.
37 "Philosophical Analysis and Historical Understanding," *Review of Metaphysics*, XXI, 1968, p. 687.
38 "Power Struggle, Parliament, and Liberty in Early Stuart England," *Journal of Modern History*, 50, 1978, p. 27.
39 *The Idea of History*, pp. 269–270.
40 "A Note on History as Narrative," *op cit.*, p. 414.
41 "Narrative Explanations in History," *Mind*, LXXVIII, 1969, p. 260.
42 *Theory and Processes of History*, Berkeley: University of California Press, 1960, p. 39.
43 *Political History*, pp. 159–160.
44 *The History Primer*, p. 171.
45 *Doing History*, Bloomington: Indiana University Press, 1971, p. 68.
46 "Narrative Form as a Cognitive Instrument," In *The Writing of History*, eds. R.H. Canary and H. Kozicki, Madison, Wis.: University of Wisconsin Press, 1978, p. 147.
47 The research upon which this paper draws was done during the tenure of a Killam Research Fellowship.

HEIDEGGER'S PHILOSOPHY OF SCIENCE: THE TWO ESSENCES OF SCIENCE

JOHN D. CAPUTO

The later Heidegger criticized the kind of thinking which, bent only on control and exploitation, threatens to overrun the contemporary technological world. He directed this critique, not precisely against technology, but against what he called the "essence" (*Wesen*) of technology, where *Wesen* had the old verbal sense of coming to be, coming to pass, coming about.[1] He was concerned, not with technological instruments themselves, but with what is coming to pass in and through a world filled with these instruments. What mattered for Heidegger is our understanding of ourselves and of the world in an age governed by the paradigm of technological control. He offered an important critique of the totalitarian tendencies of contemporary culture, in virtue of which science and technology are swept up in a repressive, totalizing movement which threatens to run out of control.

It often goes unnoticed, however, in the light of the later Heidegger's ringing pronouncements against the "essence of technology," that there is in fact "another essence" of science and technology in Heidegger's writings, an important and affirmative conception of science to be found in *Being and Time* (1927), which bears suggestive similarities with the work of Kuhn, Feyerabend, Polanyi, Hanson, Hesse and other post-positivist theories of science. Indeed, I think, the later Heidegger himself neglected his own earlier reflections, and drifted, ironically, into a positivistic conception of science.[2] But this move, I will argue, only weakened his incisive critique of contemporary culture. His own later warnings against the "essence" of technology would have been more properly served had Heidegger continued to argue, as he had in *Being and Time*, against a misunderstanding of what science is in the first place, a misunderstanding of its essence in the more straightforward sense – of what scientists do in their workshops.

Instead, driven no doubt by the imminence of the threat, Heidegger shifted his attention away from his own suggestive "hermeneutics" of science in *Being and Time*, towards the danger of a totalitarianism by which science and

Margolis, J., Krausz, M. and Burian, R.M., (eds.), Rationality, Relativism and the Human Sciences.
 ISBN 90-247-3271-9.

civilization are being subverted. And it is because of the shrillness of these later warnings that we – and that includes Heidegger himself – have tended to forget the insightful theory of science defended in *Being and Time*.

Thus I think there are in fact "two essences" of science in Heidegger, the first of which was suppressed by the second. The first, which is to be found in *Being and Time* and which we will call the "hermeneutic" essence of science, consists in a suggestive "existential genealogy" of science, its genesis in the historical life of the scientific investigator, which can not be separated from an allegedly "pure logic" of science. The second essence of science and technology is its *Wesen* in the later, more radical sense – let us call it here its "deconstructive" sense – which signifies an entire understanding of man and world, of being and truth, which Heidegger wants to delimit, to critique, to disrupt.

My aim in this paper, which turns on these two essences, is twofold. (1) I want to sketch out the main lines of the hermeneutic essence of science proposed in *Being and Time*. In so doing I hope to restore to its proper place a Heideggerian analysis which Heidegger himself tended to forget, and thereby correct a misunderstanding about Heidegger's alleged hostility to science, which Heidegger himself has in part fostered. (2) Next, I want to provide a preliminary introduction to the second essence, elaborated in his later essays on technology, and thereby put in context a critique of totalizing and totalitarian projects – which is in fact what I take the later work to be – which I regard to be of the utmost importance. And I want to show that the first sense supports the second, that the hermeneutic account serves the deconstructive critique. Supported by the hermeneutic interpretation of science defended in *Being in Time*, his later warnings about a dangerous historical momentum are not humanistic Jeremiads about science and technology, but a thoughtful protest against a growing momentum towards totalization which misunderstands science itself. In my view, the later Heidegger's protest against the "essence of technology" converges with the emancipatory and liberationist impulse which has emerged in French philosophy in the last two decades. His critique of metaphysics has a socio-political cutting edge which he himself failed to exploit, even as a deconstructive purpose is served by his earlier hermeneutic work, which he tended to forget.

I am arguing that we need both the early Heidegger and the late, both hermeneutics and the more radical deconstructive critique of metaphysics, both Gadamer (whose point of departure is *Being and Time*) and Derrida (whose point of departure is the later critique of the history of metaphysics) and recent French philosophy as a whole. I hope to show that there is a fruitful interplay, and not an antagonism, between the two essences of science. And if I cannot in fact show this within the limits of this paper, I can at least make some pointers in that direction.

An "existential conception of science"

In Section 3 of *Being and Time* Heidegger speaks of a "productive logic" which leaps ahead of the concrete work which the positive sciences do and breaks new ground, as opposed to the standard "logic":

> . . . which limps along after, investigating the status of some sciences as it chances to find it, in order to discover its 'method.'
> (SZ, 10/30)[3]

This productive logic differs in principle from a merely reproductive one, which can do nothing better than desribe the work of science in its creative past, vainly attempting to formulate that past into an epistemology by which scientific practice should presumably be guided in the future (SZ, 10/30). Such logics and methodologies are always written at dusk, to borrow Hegel's image, just when they have become obsolete. Methodologists are constantly being disrupted by the unorthodox turn of events awaiting them around the next historical bend.

We learn more about this new logic in Section 69(b), where Heidegger discusses what he calls an "existential conception of science:"

> This must be distinguished from the 'logical' conception which understands science with regard to its results and defines it as 'something established on an interconnection of true propositions – that is, propositions counted as valid.' The existential conception understands science as a way of existence and thus as a mode of Being-in-the-world, which discovers or discloses either entities or Being.
> (SZ, 357/408)

There is thus a difference between the standard logics of science, which treat science as a constituted result, attempting only to display the chain of connections among its propositions, and the existential génealogy of the origin of science from prescientific life.[4] It is the latter which interests Heidegger, and it is this shift in perspective which gives rise to a fundamentally different conception of science in *Being and Time*. The attempt to treat science as if it were a pure logic, as if it dropped from the sky, is an illusion, indeed a transcendental illusion, which vainly tries to endow science with a pure transcendental status. Science derives from the concrete historical life of the scientific investigator.

It is this new logic of science which Heidegger describes as "hermeneutic." But Heidegger's hermeneutic conception of science differs considerably from Dilthey's. In the first place, Heidegger rejects Dilthey's "objectivism," which regards science as somehow or other able to seize upon a thing in itself, whether that be nature in itself, as in the natural sciences, or indeed some cultural object in itself, as in the *Geisteswissenschaften*. Secondly, Heidegger rejected Dilthey's division of the sciences into two qualitatively different sorts, natural and human, which confines hermeneutics to but one side of the distinction. Heidegger insists that the sciences constitute a unity, albeit a hermeneutic and not a positivistic one.[5] In *Being and Time*, hermeneutics is a universal, ontological structure which included all the sciences in its sweep. "Hermeneutics" is not the name of a method of the human sciences, but of the ontology of the understanding itself. Heidegger takes hermeneutics out of the domain of the purely methodological or epistemological and shows that it ultimately has an ontological sense. We understand as we do because we are as we are. Understanding follows Being; *intelligere sequitur esse*. There is no field of pure epistemology for Heidegger but only of the ontology of knowing or understanding. Every science is a way the historical investigator has of "casting" things, of framing them out, within a certain conceptual framework. Far from attempting to leave one's own interpretive framework behind, as in Dilthey's naive objectivism, scientific understanding is not possible without it, Hence, instead of differentiating natural science from hermeneutics, Heidegger is interested in showing how all science, natural and human, is made possible by an anticipatory, hermeneutic fore-structure.

The fundamental feature of the understanding, and hence of the hermeneutic theory of science defended in *Being and Time*, is its *projective* character (*entwerfendes Verstehen*). For Heidegger, to understand is to contextualize, to situate a thing within the contextual arrangement in which it belongs. And that is what Heidegger means by "projecting" a thing "upon its horizon," or "projecting a being in its Being:" to set it forth in or on the horizonal backdrop which it requires in order to be manifest as the thing which it is, to "cast" in the appropriate terms. Indeed the English word "cast" captures a good deal of what Heidegger means by "projection." It operates within the rule of the same metaphor (*werfen* means to hurl or cast): to cast a thing in a certain light, to thrust it into a certain framework. To understand a hammer is to situate it within the chain of equipment, the equipmental context, to which it belongs, just as in the foreground-background analysis of perception in Gestalt psychology.[6] Accordingly, scientific or theoretical thinking consists in projecting an adequate conceptual horizon, one which allows, not tools or everyday things, but scientific objects to appear.

Hermeneutic understanding proceeds from a network of presuppositions which must always be adequate for the matter to be interpreted. In *Being and Time* it is never a question of getting free from presuppositions but, on the contrary, of securing an adequate presuppositional frame, of seeing to it that the complex of presuppositions one brings to bear upon the object is wide enough and sharp enough to give an adequate rendering of the object. It is never a question of assuming too much, but of assuming too little. The "hermeneutic situation," or complex of hermeneutic conditions under which understanding is possible, must accordingly meet three conditions. In the first place, the projective fore-structure must cover the whole range of entities to be understood, which is what Heidegger calls "fore-having," (*Vorhabe*). That is to say, any projective or horizonal framework which we employ múst be ample enough to provide for all the phenomena which can appear within its range. Secondly, the projective understanding must be guided by a certain conception of the kind of being, or categorial type, of the phenomena included within its range; this Heidegger calls "fore-sight" (*Vorsicht*). Finally, the preunderstanding requires an articulate table of categories which unfold, delineate, and provide for an adequate analysis of, the mode of being caught sight of in fore-sight; this is called "fore-grasping" (*Vorgriff*) (cf. SZ, 150–51/191–2, 231–33/274–5).

Together these conditions supply the hermeneutic forestructures, the anticipatory conditions of possibility, the required presuppositions which constitute the pre-understanding under which explicit understanding is possible. *Being and Time* clearly defends a holistic, horizonal interpretation of understanding according to which the character of an individual act of understanding is set by the constellation of presuppositions which condition it and make it possible. These presuppositions hang together in a system so that there are no isolated, atomic acts of understanding, on the one hand, nor non-contextualized objects on the other. It is just this projective, horizonal theory of understanding that we need to characterize more carefully in order to grasp Heidegger's theory of science in *Being and Time*.

In the first place, the hermeneutic forestructuring of human understanding is an ontological condition rooted ultimately, according to Heidegger, in the temporality of human existence as a being pointed towards the future. That is why talk about "escaping" this fore-structuring – which is what "freedom from presuppositions" would mean – makes no sense. On the one hand such "escape" would be ontologically impossible; one would have to escape from one's condition as a temporal being. And, on the other hand, such talk of escape misunderstands the very nature of understanding which proceeds, not by means of a presuppositionless blank stare, but by projecting the horizonal preunderstanding which befits a given category of entities (SZ, 153/194–5).

Accordingly, the hermeneutic make-up of the sciences is but a particular instance of the universal hermeneutic structure of all understanding. The Diltheyan opposition of explanatory-causal natural sciences and clarifying-hermeneutic human sciences makes no sense in *Being and Time*. Both explaining and clarifying, both natural and human science, are possible only on the basis of a prior hermeneutic projection which constitutes the field of objects carved out by that discipline. Heidegger shows little interest in, and puts little emphasis upon, the distinction between the natural and the human sciences in *Being and Time*. Indeed, he takes all the sciences to have the same ontological weight. All science, as science, is a projective determination of beings in terms of some categorial framework or another – be it natural, social or human. What does interest Heidegger, however, is a different and, for him, far more important distinction – between science itself and prescientific life. It is that distinction which his existential genealogy seeks to establish and clarify; it is upon that distinction that he thinks everything depends.

Let us examine this genealogy more carefully. *Being and Time* emphasizes the primacy of our concrete involvement in the sphere of historical and cultural practices and the secondary or derivative status of abstract, theoretical investigations. The concrete historical world is first of all the world of "instruments" whose primary character is their "being ready to hand" (*Zuhandensein*), that is, their availability for use. *Zuhandensein* is to be differentiated from the complementary concept, *Vorhandensein*, which refers to the world in its "objective presence," its sheer reality apart from our use, which is the way it is considered in the natural sciences. However, we must resist the temptation to give the theoretical-scientific relationship to the world a purely negative genealogy, as if it arises as the simple privation of our primary practical engagement with the world (= "circumspective concern"):

> . . . it would be easy to suggest that merely looking at entities is something which emerges when concern *holds back* from any kind of manipulation. What is decisive in the 'emergence' of the theoretical attitude would then lie in the *disappearance* of *praxis*.
> (SZ, 357/409)

That would flatly contradict the universally projective nature of understanding, suggesting that, while our concrete involvement with the world is hermeneutic and presupposition-bound, the natural sciences treat the world in an objective, presuppositionless way.

Consider the case of the hammer. The hammer which is used *as* a hammer is grasped only in the using; one cannot explain in theoretical terms its "feel," its balance, its aptness as an instrument, its place in the instrumental system. One knows how to use a hammer only by trying his hand at hammering with it. But one can step back from using the hammer and descibe its "properties" in a series of "assertions," indicating for example that "the hammer is heavy." In that case the hammer shows itself in a new light, but not merely because we no longer use it:

> Not because we are keeping our distance from manipulation, nor because we

> are just looking *away* (*absehen*) from the equipmental character of this entity, but rather because we are looking *at* (*ansehen*) the ready-to-hand thing which we encounter, and looking at it 'in a new way,' as something present-at-hand. The *understanding of Being* by which our concernful dealings with entities within-the-world have been guided *has changed over*. (SZ, 361/412)

As an object of the theoretical attitude, the hammer ceases to be a hammer-in-use and is *recast* as a thing-with-properties, obedient to the laws of gravity, measurable in mathematical space, etc. It is projected positively and anew, literally re-cast, re-projected, now no longer on the horizon of its readiness-at-hand, but rather of presence-to-hand. (One can imagine a graded series of such assertions, from the after-hours talk of two carpenters comparing the objective properties of various sorts of hammers, all the way to the considerations of a physicist who would treat it purely in terms of its mass and velocity.)

Hence it is a mistake to think that in the scientific attitude things lose their interpreted character, their character *as* something (which Heidegger calls their "as-structure"). It is a mistake to think that the ready-to-hand is treated *as* a tool while the present-at-hand has to do with things *in themselves*, free of any "as." On the contrary, we have simply switched from one "as" to another (SZ, 157–8/200–201). We have simply shifted projective frameworks, hermeneutic forestructures. We have ceased to regard the hammer as a tool and now regard it as an entity with mass, shape, gravity, etc. In other words, "objective presence" is a positive projection of the Being of entities, a hermeneutic-interpretive act. (It should also be clear from this passage that there is nothing mystical or mystifying about the word "Being" as Heidegger uses it, particularly in *Being and Time*; it simply refers to the conceptual framework, the horizon, in terms of which a thing is grasped [= "projected"]).

Furthermore, one ought not to treat the genesis of the scientific attitude from our concrete involvement with the world as a passage from practice to theory. For that presupposes a hard and fast distinction between practice and theory – as if theory did not have a praxis of its own, and practice did not have a 'sighting' of its own – a distinction which Heidegger thinks cannot be defended. On the contrary, scientific work is a complex praxis which requires skilled investigators, and this is no incidental feature of science, but of the utmost importance to it:

> Reading off the measurements which result from an experiment often requires a complicated 'technical' set-up for the experimental design, Observation with a microscope is dependent upon the production of 'preparations'. Archaeological excavation, which precedes any interpetation of the 'findings', demands manipulations of the grossest kind.

> But even in the 'most abstract' way of working out problems and establishing what has been obtained, one manipulates equipment for writing, for example. However 'uninteresting' and 'obvious' such components of scientific research may be, they are by no means a matter of indifference ontologically. The explicit suggestion that scientific behavior as a way of Being-in-the-world, is not just a 'purely intellectual activity', may seem petty and superfluous. If only it were not plain from this triviality that it is by no means patent where the ontological boundary between 'theoretical' and 'atheoretical' really runs!
> (SZ, 358/409)

It is also important to see that Heidegger does not think that science is exhausted by, or restricted to, the change-over in conceptual frameworks from readiness-to-hand to objective presence. For that change-over explains the possibility of the natural sciences, but not of the social and humanistic sciences. Science is constituted, not only when cultural objects (tools) are treated as physical objects, but also when tools are grasped *in their tool-ness* and made objects of scientific inquiry:

> The ready-to-hand can become the 'Object' of a science without having to lose its character as equipment.
> (SZ, 361/413)

Hence economics is a science of the goods men make, and the laws of supply and demand by which they are regulated, which preserves the character of the tool as a tool. Economics is a sciences of the ready-to-hand *as* ready-to-hand.

The genealogy of the scientific attitude in *Being and Time* therefore involves two things: (1) the change-over from the attitude of immediate involvement ("concerningful dealing") to a certain distancing objectivity; (2) the positive projection of a kind of Being which befits the beings under investigation, which will supply a horizonal or hermeneutic framework within which they can be investigated. In the natural sciences, that means the projection of things in terms of their objective presence, but in the social and human sciences it obviously does not. Economics, politics, literary and historical science would cast things in their objective presence only at the cost of their existence as meaningful inquiry. The natural and human sciences arise from a different sort of projective horizon, with different conceptions of time, space, law and meaning. These differences amount, however, to "ontic" differences in *Being and Time*, inter-disciplinary differences about their presuppositions (their fore-having, fore-sight and fore-grasping). They do not affect the ontological make-up of science as science, but the intra-scientific differentiation of the sciences among themselves.[7]

Heidegger's concern in *Being and Time* is to see to it that the natural, social and human sciences – which together are characterized as "possitive" or "ontic" sciences, for they have to do with entities, with posited, existing things – do not get mixed in, indiscriminately, with the ontological science which he is conducting in this work. His ultimate interest in *Being and Time* is in Being, not beings, that is, in the ontological framework which renders possible every human practice, both scientific and pre-scientific, or, as he puts it, every ontic comportment with beings (*Verhalten zu Seiendem*, SZ, 4/23). The aim of *Being and Time* is to reach a determination of Being as the horizon, not of any particular region, but of all horizons, and this he will argue is "time." It was thus of the utmost importance that the inquiry into "Dasein," which serves as the point of departure in this ontological investigation, not be taken in ontic terms, as if it were an anthropological, psychological, or sociological science. And while he wished to keep this purely ontological interest in man in *Being and Time* distinct from the positive-ontic sciences of man, it was not his intention to question the validity of the latter. He wanted neither to demean the natural sciences in the light of the human sciences, nor to demean the particular, positive sciences in the light of fundamental ontology. On the contrary, he meant to offer a hermeneutic account of the possibility of science itself, and to offer an ontology of man as a hermeneutic being.

The genesis of science is to be located, not in its object, but by the *projective attitude* with which it carves out and constitutes a field of objects and defines a formal approach to them. When an historian trains his student to look for certain patterns, to value certain documents, to be on the alert for certain clues, he is engaged in establishing the projective standpoint which constitutes and defines a scientific inquiry into history. And the same thing is true of the literary critic teaching his students to thematize poetic imagery or the construction of English novels in the nineteenth century, or of a physicist working with his assistants in the laboratory.

Science is objectification, and objectification, which is the constitution of any sort of scientific object, whether natural, social or human, is effected by the change-over (*Umschlagen*), a new way of projecting beings in their Being. The piece of equipment is released from its place within the equipmental totality and treated as a detachable thing, isolated from its equipmental context. But if it is decontextualized as equipment it is recontextualized as an object within a new theoretical framework and made the object of a new and more sophisticated praxis. It makes its appearance as a scientific object only by assuming its place within a new horizonal setting.[8]

It is also clear from this hermeneutic or projective theory of science, that Heidegger is committed to denying "bare" or uninterpreted facts of the matter. Indeed, for Heidegger, the selection of facts in any science is a function of its capacity to discover a way to project things; a scientist can pick out facts only in virtue of a prior frame which he "has in advance." Take the case of physics, which is for Heidegger a "paradigmatic" or exemplary case of scientific projection. Mathematical physics proceeds by projecting a strictly quantitative nature, a nature cast or projected in terms of mass, location, velocity, etc. which thereby enables it discover facts:

> The 'grounding' of 'factual science' was possible only because the researchers understood that in principle there are no 'bare facts.' In the mathematical projection of nature, moveover, what is decisive is not primarily the mathematics as such; what is decisive is that this projection *discloses something that is priori*. Thus the paradigmatic character of mathematical natural science does not lie in its exactitude or in the fact that is binding for 'Everyman.' It consists rather in the fact that the entities which it takes as its theme are discovered in it in the only way in which entities can be discovered – by the prior projection of their state of Being.
> (SZ, 363/414)

This is one of the most significant points of contact between Heidegger and the recent rereading of the history and philosophy of science. There are no facts except within the pre-given horizon which enables them to appear in the first place.

However, one ought not to conclude from Heidegger's denial of bare facts that he regards all projective horizons as arbitrary or pragmatic fictions. On the contrary, projection has for him a "disclosive" power. Understanding discloses the world as the sort of world it is; it renders it manifest in a certain way. And disclosure, manifestness, is what "truth" means in *Being and Time* (SZ, sec. 44c). Projections are not arbitrary; we must involve just the right kind of framework to free beings up for the kind of Being which befits them. The sciences do not traffic in "free-floating constructions" (SZ, 28/50) but in fact seize upon something in the things themselves. Thus Heidegger holds *both* that it is impossible to gain access to bare and uninterpreted facts of the matter – which is to reject any notion of objectivism or absolutism, as if we could jump out of our skins and make some absolute contact with things – *and* that our hermeneutic constructions, when they are well-formed, do capture something about the world – which is to provide for the objectivity of knowledge. Thus in one stroke Heidegger provides for the possibility of science while delimiting the claims of objectivism. He has no reason to think that science does not seize something about the world, even while he thinks it nonsense to suggest that this constitutes a break in the hermeneutic circle.

Scientific crises

Up to this point we have treated Heidegger's views on what Kuhn would call normal science. However, there is also a clear picture of "revolutionary" science in *Being and Time*, in Section 3, where Heidegger treats of the "crises" and "radical revisions" of "fundamental concepts" which periodically shake the sciences. The sciences, as we have seen, are guided in advance by certain basic concepts (*Grundbegriffe*), the interpretive fore-structures that constitute or project the field of objects that belong to that science:

> Basic concepts determine the way in which we get an understanding beforehand of the area of subject-matter underlying all the objects a science takes as its theme, and all positive investigation is guided by this understanding.
> (SZ, 10/30)

"Basic concepts" are a good deal like Kuhnian paradigms – guiding fore-structures which guide a whole scientific practice. The difference is that speaking of "concepts" tends to obscure Kuhn's emphasis on the practical character of the paradigm, an emphasis, however, which Heidegger clearly would share (SZ, 358/409). Now just as the transition from everyday concern to science is effected by a fundamental change-over (*Umschlagen*), which is a shift from a prescientific to a scientific projection, so *within* each science certain fundamental shifts of projective understanding are possible which result in revolutionary changes within that science:

> The real 'movement' of the sciences takes places when their basic concepts undergo a more or less radical revision which is transparent to itself. The level which a science has reached is determined by how far it is *capable* of a crisis in its basic concepts.
> (SZ, 9/29)

At the time he was writing *Being and Time*, Heidegger thought that any number of such fundamental revolutionary movements were underway. He speaks of a crisis in mathematics, in the dispute between formalists and intuitionists; in physics, as a result of relativity theory; in biology, because of the dispute between vitalism and mechanism; and in theology, because of Luther's fundamental insights about the nature of faith. So the work of a Galileo, Newton or Einstein, of an Aquinas or Luther, of a Smith or a Keynes, or any of the founding or revolutionary geniuses of the respective disciplines, consists not so much in making new factual discoveries, or generating new information, as in effecting certain fundamental conceptual breakthroughs, in revising radically the fundamental terms in which the practitioners of a discipline think about a field of objects, in thoroughly "re-casting" them.

Inasmuch as Heidegger holds to a "horizonal" or holistic theory of understanding, he regards progress in a science as possible on two levels. In the first place, the scientist can continue to fill in the existing horizon, building up in a continuous way the known body of information (confirming predictions, refining calculations, etc.). This is what Husserl called the "fulfillment" (*Erfüllung*) of a predelineated horizonal scheme (*Logical Investigations*, Investigation VI, Section One, Chapter 1). And it is also, of course, the everyday business of what Kuhn called normal science. But it is possible – and sometimes necessary – that the horizon itself undergo revision. and that can occur only by a discontinuous revision or shift of horizons. Certain fundamental thinkers in a discipline, working at its boundaries, force a reorganization of the whole field of disciplinary activity. In the language of phenomenology, their work is carried out on the level of "regional ontology," that is, of the ontologically guiding and horizonal concepts within which all the work in their field is conducted.[9] The great creative geniuses work on and at the horizons within which their more pedestrian colleagues labor unquestioningly. The phenomenology of horizons explains the phenomenon of scientific revolutions.

The history of science is punctuated by these horizonal shifts which reorganize the data contained within the horizon. How, then, are we to think of scientific "progress"? Heidegger wrote in 1938, over two decades before Kuhn:

> When we use the word 'science' today, it means something essentially different fromm the *doctrina* and *scientia* of the Middle Ages, and also from the Greek *episteme*. Greek science was never exact, precisely because, in keeping with its essence, it could not be exact and did not need to be exact. Hence it makes no sense whatever to suppose that modern science is more exact than that of antiquity. Neither can we say thay the Gallilean doctrine of freely falling bodies is true and that Aristotle's teaching, that light bodies strive upwards, is false; for the Greek understanding of the essence of body and place and of the relation between the two rests upon different interpretations of beings and hence conditions a correspondingly different kind of seeing and questioning of natural events. No one would presume to maintain that Shakespeare's poetry is more advanced than that of Aeschylus. It is still more impossible to say that the modern understanding of whatever is, is more correct that that of the Greeks. Therefore, if we want to grasp the essence of modern science, we must first free ourselves from the habit of comparing the new science with the old solely in terms of degree, from the point of view of progress.[10]

The individual scientist works within a projective horizon which sets forth its own standards of what is reasonable and scientific. Scientific theories belonging to different projective structures, operating within different constellations of basic concepts, may be compared and contrasted as to their presuppositions, but one cannot be labelled more rational or progressive than the other.

On an horizonal theory of the understanding, therefore, scientific development is not uniform and continuous, but at certain times must be shocked by more or less radical revisions of the existing horizons. But what in particular occasions such crises? What are the reasons which lead scientists to abandon the old horizon and adopt a new one? What place can the data gained under the old horizon assume within the new horizonal framework? Heidegger does not answer, or even ask, these questions in *Being and Time*. That is why I think that Kuhn's conception of scientific revolutions is not only congenial to the standpoint of *Being and Time*, but in fact elucidates, works out and corrects what is only a seminal suggestion in Heidegger. Indeed Kuhn's approach is so preeminently in keeping with the hermeneutic conception of science which we are defending here that he himself has been led to describe his view as "hermeneutic."[11]

If Heidegger failed to investigate the character of the decision which the scientist makes at the point of crisis, when the old horizon wavers in instability and a basic shift is about to be made, Kuhn's account of this moment created a storm of criticism, including Lakatos' famous observation that Kuhn reduced scientific decision making to mob-psychology.[12] Kuhn might have avoided this criticism had he at his disposal the hermeneutic conception of *phronesis* which Gadamer developed by listening to the young Heidegger's lectures on the *Nichomachean Ethics* and whose counterpart in *Being and Time* is the theory of *Verstehen*.[13] The Aristotelian model shows that intelligence does not always have explicit rules to fall back upon, that at certain critical points it is left to its own devices to grasp what the situation demands. This does not mean that for Aristotle we are sometimes driven to act irrationally. On the contrary, the understanding is never more faithful to its nature than it is in these moments. For it is precisely the work of understanding to make the first cut into the complexity of the concrete world, to find the nerve of intelligibility which runs throught it. It is precisely at these moments, when it lacks rules to fall back upon, that intelligence must be what it most essentially is: insightful, capable of grasping what is demanded in a concrete setting. It is to Gadamer's credit to have elaborated the notion of *phronesis* in an admirable discussion in *Truth and Method*,[14] and in so doing to have set forth an essential implication, not only of Aristotle, but of *Being and Time*, for the theory of the sciences.

We should not fail to notice, moreover, that Heidegger's examples of crises in fundamental concepts in the sciences cut across the spectrum of natural, social and human sciences, including in one sweep; physics, economics and theology. Such revolutions affect *Wissenschaft* itself,[15] in the wide sense that this term has in German, and which for Heidegger means any disciplined investigation, in which a particular region or field of objects has been staked off and thematized (objectified) in terms of a certain projection or conceptual organization. That happens both in theology and physics, whose respective histories are punctuated by basic upheavals in their forestructures. Accordingly, one requires *phronesis* – hermeneutic insight and understanding, a feel for what one is about, which Polanyi calls "personal knowledge" – across the board: in the natural, social and human sciences.

***

It is clearly a myth to think that there is a hostility or denigration of the sciences in *Being and Time*. On the contrary, Heidegger brings to fruition in this book the work he began as a student of Rickert who, as a leader of the Baden school of Neokantianism, had written extensively on the theory of the sciences, a point clearly reflected in Heidegger's *Antrittsrede* at Freiburg in 1915.[16] Heidegger is interested, not in undermining the sciences, but in providing them with a hermeneutic accounting. In particular, he wants to show that scientific activity, of whatever disciplinary type – natural, social or human – is nourished by a prescientific, historical life which is its matrix and point of departure. He wants to explain how science is "derived" from historical life (= "Being-in-the-world"), how it is ontologically generated from our concrete entanglement with the world – viz., by a horizonal change-over from our primary and inescable "concern" with the world to a relatively disengaged projection of the world as a field of objects. He wants to show that, while legitimate in its own sphere, any such scientific projection is limited. Scientific projections are theoretical constructions aimed at elucidating a world from which we cannot finally or wholly extricate ourselves and to which we belong more primoridally than science itself can say. It is always that prescientific belonging to the world which has primacy for Heidegger. It is that world which Heidegger ultimately wants to elucidate. In so doing he does not intend to hold science in contempt, but only in check. He delimits its claims by subordinating science to the world in which we live and which has a prior claim upon us. If science is made possible by a hermeneutic projection of a sphere of objects, it is also limited by the hermeneutic horizon of the scientist. A *pure* scientific standpoint is thus an illusions; science is always a projective undertaking of Being-in-the-world.

The "essence of technology" in the later Heidegger

Heidegger's later writings on science and technology are marked by a dramatic change of tone.[17] The sober hermeneutic analyses, suffused with the technical vocabulary of phenomenology, give way to the voice of protest. The determination of the first, hermeneutic essence of science is displaced by a deconstructive critique of the *Wesen* of technology, of what is coming to pass, what is coming about, in a technological world. There is a tone of urgency and of shrill protest against a danger which he takes to be imminent. He is no longer concerned with a logic of science – not even an existential or hermeneutic logic, a genea-logic. Science belongs together with technology as inseparable forms of the will-to-power as knowledge, of the will-to-know, the will to dominate and manipulate, of what Foucault calls "power-knowledge." The metaphysics of the will-to-power stamps our age, marks our epoch, dominating all the phenomena of our time – political and social, scientific and artistic. We are in danger of being swept up in an enormous totalitarian and totalizing movement which aims to bring every individual, every institution, every human practice under its sway.

His protest, as we have said, is directed, not against science and technology themselves, but against their "essence," which means, what is coming to pass in a world that gives science and technology paradigmatic status. He is not concerned with science and technology, but with the way understanding being, man and truth their enormous success and growing prestige have induced in us, an understanding which he thinks is marked by a preoccupation with control, manipulation and power. The metaphysics that articulates the guiding conceptions of modern technological civilization is to be found in Nietzsche's conception of the will-to-power. A dangerous momentum has been set loose in the modern world, one that conceives nature, and man himself, as the raw material of a manipulative technology, that conceives all problems – political, social, personal – as technological problems for which an appropriate technology of behavior is required. The world has become the raw material for the various technologies of power – political technologies that manipulate and control public opinion and policy; social technologies that manipulate and control personal mores and standards of conduct; educational technologies that insure the normalization and regulation of educational practices. It is not only nature which must submit to our control, but education, sexuality, the political process, the arts, in short the whole sphere of human practices. This is what is coming to pass (*Wesen*) in science and technology, not in the sense that it is explicitly taught by science, but in the sense that this is the frame of mind of a culture dominated by the success and prestige of science and technology.

We are caught up in a momentum which is running out of control, and

Heidegger's later work is a philosophy of liberation, emancipation, protest against that momentum. It has, I think, a profound kinship with the work of Derrida and Foucault in France, with Adorno's negative dialectics – despite the latter's well-known diatribes against Heidegger – in German social theory, and in the United States perhaps with Feyerabend. It is an eloquent protest against normalization, regulation, manipulation, against the rule of the police. It is a philosophy of freedom and letting-be, which wants to let the world, to let man, to let the gods, be.

Unfortunately, in the course of lodging a protest which could not be more salutary against what is coming to pass in and with science, against its *Wesen* in this special sense, Heidegger loses sight of his own more careful, more perspicuous characterization of science itself – its "essence" in the first, ordinary or hermeneutic sense. Thus in his later writings Heidegger speaks of science as if it were an unbending method which knows only how to apply fixed rules to unchanging circumstances, as if it were, by its nature, by its first essence, part and parcel of the disciplinary society. Unhappily, he seems to forget what he himself said in *Being and Time* about understanding as *phronesis*, about the historicity of the scientific investigator, about horizonal shifts and breakthroughs, crises and ground-breaking discoveries. And that is a regrettable development. For his argument is with the totalitarian forces which exploit science and which sweep it up in a vast political and economic armature, a complex of forces and power which knows no bounds. That argument would have been strengthened, not weakened, had he continued to insist, as he did in *Being and Time*, that science itself, in its real practice and practical reality, in its first, hermeneutic essence, is not the rigorously rule-governed, inflexible apparatus it is made out to be. That is why we need both the early Heidegger and the later, or among contemporary post-Heideggerian figures, both Gadamer and Derrida. We need both a positive hermeneutic rendering of science such as we find in Gadamer, who represents something of a right-wing, conservative, early-Heideggerianism, and a vigorous protest against standardization, normalization, and ideological imperialism, such as we find in Derrida, who represents a left-wing Heideggerianism, and in Foucault and recent French philosophy.

Despite this slip in strategy in the later Heidegger's reflections on science – and perhaps, in a philosophy of protest, moderate, balanced accounts are not always what we need! – one can see, nonetheless, the unity of Heidegger's project from *Being and Time* on. He has been concerned throughout with that world which precedes science, which funds it, to which science always returns, and from which the scientist himself is never granted leave. In *Being and Time* this prescientific sphere was characterized in terms of the historical, cultural "life-world" of phenomenology, and in the later writings in terms of a more poetically conceived life of "mortals," "under the skies," "before the gods," "upon the earth," a picture inspired ultimately by the poetry of Hölderlin. It is as if the threat had grown so great that we required a more soaring vision, a more powerful mythology, than was provided by the sober analytics of Husserlian phenomenology. But he meant all along to recover the prescientific world which antedates science and which has its hold upon us long before science arrives on the scene, and his point has always been to *let* that world *be*.

NOTES

1 *Wesen* was used as a verb in Middle High German; whence the modern *west*. The English translators of Heidegger tend to render it as "come to presence."

2 This point runs throughout the work of Theodore Kisiel, who has done the best work on Heidegger's philosophy of science. See his "Heidegger and the New Images of Science," *Research in Phenomenology* 7 (1977), 162–81; "New Philosophies of Science in the USA: A Selective Survey," *Zeitschrift für allgemeine Wissenschaftstheorie* V (1974), 138–91; also see note 4 below. On Heidegger and Polanyi, see Robert Innis, "Heidegger's Model of Subjectivity: A Polanyi Critique," in *Heidegger: The Man and the Thinker*, ed. Thomas Sheehan (Chicago: Precedent Publishing Co., 1981), 117–30.

3 SZ = Heidegger, *Sein und Zeit*, 10 Aufl. (Tübingen: Niemeyer, 1963); the pages following the slash are to *Being and Time*, trans. E. Macquarrie and J. Robinson (New York: Harper & Row, 1962).

4 As in the current discussion of what Popper calls the "logic of discovery," Heidegger thinks that the mistake of the philosophy of science up to now has been to concentrate on the finished results of science, on science as a body of established propositions, rather than on the process by which such propositions arise. But the danger entailed by the genealogical approach – which wants to avoid being a pure logic – is psychologism and relativism, as Husserl pointed out in the first volume of the *Logical Investigations*. As an application of phenomenology, Heidegger's hermeneutic genealogy intended to steer a middle course between the two. As a hermeneutic, it is no pure logic, but historical and genetic; as a phenomenological ontology, it is concerned, not with human psychology, but with the ontological structure of understanding [of "Dasein," not of "man."] The disagreement between Popper and Kuhn in *Criticism and the Growth of Knowledge* [ed. Imre Lakatos and Alan Musgrave (Cambridge: University Press, 1970)] is illustrative. Kuhn takes Popper's concerns to be too strongly dictated by an ahistorical logic (pp. 21–22) and Popper takes Kuhn's views to be relativistic and psychologistic (55–58). Kuhn would have done better in responding to the charge of psychologism to argue that his work is hermeneutic. To argue, however, that it is "social" and not "individual" psychology does not answer Popper's charge but simply confirms and refines it. Theodore Kisiel makes this important point in his "Scientific Discovery: Logical, Psychological or Hermeneutical?" *Explorations in Phenomenology*, ed. David Carr and Edward Casey (The Hague: Nijhoff, 1973), 263–84; See also his "The Rationality of Scientific Discovery" in *Rationality Today/La Rationalité Aujourdhui*, ed. Theodore Geraets (Ottawa: University Press, 1977), pp. 401–11.

5 If the positivists supported a "unity of the sciences" program inasmuch as they wanted to reduce all the sciences to the method of the natural sciences, which was the very thing Dilthey opposed, Heidegger treats all science as a hermeneutic unity, i.e., an exercise in projective understanding. That is not to reduce all science to the method of one privileged science but rather to make all sciences generally conform to the ontological-hermeneutic structure of the understanding.

6 Gestalt psychology is a common point of reference for Kuhn, Polanyi, Merleau-Ponty, Husserl and Heidegger.

7 It is important for us to observe that Heidegger never questions the legitimacy of the social and human sciences, or that they play a valid role in the work of *Wissenschaft* as a whole.

8 It is not clear to me that Hubert Dreyfus takes this into account in his various treatments of Heidegger's philosophy of science in *Being and Time*. He seems to me overly fond of emphasizing the decontextualization of the ready-to-hand without insisting on the concomitant recontextualization or hermeneutic projection without which understanding is impossible for Heidegger.

9 Heidegger seems to think that such fundamental conceptual breakthroughs would be effected by regional ontologists; this makes sense so long it is recognized that the revolutionary figures in the disciplines themselves are their own regional ontologists. For Kuhn, they are made by people working at the most advanced and specialized level of puzzle-solving in that discipline.

10 Heidegger, *The Question Concerning Technology and Other Essays*, trans. W. Lovitt (New York: Harper & Row, 1977), pp. 117–18.

11 Kuhn seems to have in mind, however, only the Diltheyan sense of hermeneutics as historical empathy. He does not at all intend the Gadamerian view that scientific understanding depends upon *phronesis*. See *The Essential Tension* (Chicago: University Press, 1977), pp. xiii, xv.

12 Imre Lakatos, "Falsification and the Methodology of Scientific Research Programmes," in *Criticism and the Growth of Knowledge*, p. 178. This volume contains a series of mostly critical responses to Kuhn; cf. the contributions by Popper, Toulmin and Watkins. Feyerabend, on the other hand, "defends" Kuhn by saying that science is indeed at least (and in fact even more) irrational than Khun holds, a defense which Kuhn describes as "vaguely obscene" (264) in his instructive "Reflections on my Critics" at the end of the volume.

13 Hans-Georg Gadamer, *Philosophical Hermeneutics*, trans. David Linge (Berkeley: Univ. of California Press, 1976), pp. 201–2. I have learned a great deal about a rapprochement between Kuhn and Gadamer from Richard Bernstein's insightful application of hermeneutics to the problems of the philosophy of science in *Beyond Objectivism and Relativism* (Philadelphia: University of Pennsylvania Press, 1984).

14 Hans-Georg Gadamer, *Truth and Method*, trans. G. Barden and J. Cumming (New York: Seabury, 1975), pp. 274–305.

15 The English word which seems to me to cover the range of all the sciences in the manner of the German "*Wissenschaft*" is "discipline." This word, of course, has a Foucaultian ring which, while foreign to *Being and Time*, would be a welcome nuance in the later Heidegger.

16 Heidegger's *Antrittsrede* at Freiburg in 1915, entitled "The Concept of Time in the Science of History," differentiated the historian's and the physicist's conceptions of time. See *Frühe Schriften* (Frankfurt: Klostermann, 1972), 413 ff. and my review "Logic, Language and Time," *Research in Phenomenology* 3 (1973), 147–56.

17 See the first two essays in *The Question Concerning Technology and Other Essays*.

Part Two

THE INTELLIGIBILITY OF ACTION*

ALASDAIR MACINTYRE

When I play chess with my son, my son plays chess with me. When I play chess with my computer, my computer does not play chess with me. And this in spite of the fact that my computer is a more skilful chess-player than either my son or I. Some have supposed that the only ground which would warrant such an assertion would be a compelling argument to the effect that there are specific types of task which human beings can perform, but machines cannot. But this seems to me to be mistaken and in any case it is clear that we possess no such compelling arguments. Others have claimed that it is because machines must lack anything adequately resembling human consciousness that we are not entitled to understand their performances as actions. For reasons that will emerge much later in this paper this too seems to me to be mistaken. What I shall claim is that characteristically human actions have the property of intelligibility, although some human actions may lack that property without thereby ceasing to be human actions; whereas the performances of machines are neither intelligible nor unintelligible. What then do I mean by intelligibility?

I.

I smile in a friendly way at someone at a party. He responds by saying "Have we met somewhere before?" My opponent at chess takes my queen. I say "I might as well resign. Do you want another game or shall we go for a walk?" We go for a walk. A cook tastes his fish stew and says "A little more pepper, perhaps. What do you think?" His friend takes the spoon, tastes the stew and nods. In each case the response is an intelligible sequel to the immediately prior action or actions and as such presupposes an interpretation of that behavior to which it is a response as an action of such and such an intelligible kind. Were we unable to evoke such responses we would be unable to participate in those networks of transactions within which the actions of individual agents find their place. Hence it has to be an essential part of learning to act that we learn

* I am indebted for penetrating and constructive criticisms of an earlier version of this paper to Amélie Oksenberg Rorty, Wesley Felix and especially Joseph Margolis.

Margolis, J., Krausz, M. and Burian, R.M., (eds.), Rationality, Relativism and the Human Sciences.
 ISBN 90-247-3271-9.

to behave in such a way that others construe our actions as intelligible. For otherwise our behaviour would in the eyes of others have a surd and unintelligible character; as indeed it sometimes does. But when it does, we are in danger of being treated literally as idiots.

Intelligibility is however not only a property which our actions must possess if they are to be understood by others as actions, to which response may then be made by further actions on their part. We ourselves have to be able to understand our own actions as intelligible, perhaps warranted as a response to some action by another and certainly warranting some kinds of response rather than others, if we are to be able to acknowledge them as actions rather than as something which we simply find ourselves doing, being as puzzled as anyone else as to what we are up to in doing whatever it is.

Intelligibility is a property of actions in their relationship to the sequences within which they occur. Sometimes this is a matter of an immediate temporal relationship. Given what has just occurred to someone, for that person to behave in some ways is intelligible, for them to behave in certain other ways unintelligible. But sometimes that to which an action is either an intelligible or an unintelligible sequel may have occurred many years previously. In Book VI of the *Iliad* the conduct of Glaukos and Diomēdēs towards each other on the battlefield is made intelligible by reference to a meeting of their grandfathers many years before. Sequences are thus of different types: conversations, feuds, enquiries such as those of the sciences, projects in the arts or politics, playing through a particular game and indefinitely many more long- or short-term individual or joint types of project and transaction.

To learn to act then is one and the same as learning to participate in such sequences. We do often enough as small children – and occasionally reenact this experience when as adults we enter largely alien cultures – by first having a particular meaning imputed to our behaviour and thus coming to understand our own behaviour as behaviour having this specific kind of meaning for others. We become what others already took us to be. But although we can only become the authors of intelligible action through making our own a kind of understanding and a certain set of judgments which were originally imposed upon us by others, it is also true that once we are participants in the sequences of intelligible action, our actions may well *seem* unintelligible to others, but actually *be* intelligible, or indeed *vice versa*. Intelligibility is an objective property of actions or of sets of actions; it is not in the eye of the beholder.

In cases in which an action is intelligible, but seems unintelligible to observers, it will of course generally be the case that there is some feature of the action in its relationship to the sequences of which it forms a part of which those observers are unaware. One cause of this may be that some crucial stage or stages in one or more of these sequences has occurred within the mind of the agent as a set of thoughts and also perhaps of feelings. So an observer may have witnessed an action on the part of X towards Y which renders a certain set of responses by Y appropriate and when Y acts towards X in some quite other way finds Y's actions unintelligible. What the observer did not and could not have witnessed in such a case were Y's intervening reflections, perhaps Y's remembering and reflecting upon some relevant past episode in Y's relationship to X so that it is to that, instead of to or as well as to what X has just done, that Y responds. And if and insofar as those reflections occasion Y's subsequent actions, what it is about those actions that make them intelligible may remain obscure to the observer and indeed to X. Sequences of action pass through the private world of the mental as well as the public world of the social.

For this to be the case there must be relationships of cause and effect connecting the realms of the private and the public, so that doings and happenings in the public world may be the causes of doings and happenings in the private world and vice versa. And within the private world also cause and effect relationships must hold if there are to be extended chains of private occurrences linking public occurrences. So the application of the concept of intelligiblity to actions presupposes the applicability of some conception of the causation of states and events compatible with our understanding of intelligibility. But before we can fruitfully enquire what this presupposition amounts to, it is important to notice some other features of intelligibility.

An unintelligible piece of behaviour may nonetheless be an action. That is to say, it may be informed by intention and be performed deliberately and voluntarily. But it will be able to provoke in others only some kind of baffled response and the agent him- or herself will only be able to give a very limited account of what he or she takes or took him- or herself to be doing or have been doing in performing it. It is therefore not a sufficient condition for an action or a set of actions to be intelligible that it or they should be intentional – in a sense that excludes what is done only by accident or through ignorance, so that what I actually do does not implement or embody my intention, but it is of course a necessary condition. And I can therefore rebut an interpretation of my actions as warranting a certain kind of response insofar as I can show that those actions did not in fact implement or embody the intention ascribed by that interpretation.

That unintelligible actions may occur is not only of course an abstract possibility. The occurrence of the brutely unintelligible is indeed a recurrent feature of human actions and transactions. But the characterisation of that unintelligibility and the significance to be attached to it depend upon the nature of the context of intelligible action and transaction within which the particular episode of surd unintelligibility occurs. The unintelligible is thus a phenomenon secondary to the intelligible and the understanding of human action requires from the outset an adequate account of intelligibility. In virtue of what then do we treat a particular action or set of actions as intelligible? The answer is: in virtue of its or their relationship to certain kinds of social institution and practice.

II.

The primary form of institutionalised social setting required for actions to be intelligible – in normal circumstances – is that provided by whatever the established routines are which in a particular social group constitute the structure of the normal day. In a normal day meals are prepared and eaten at set times and appointed places, schedules for times of work and leisure are observed, routine duties prescribed for this day of the week or the month or the year are performed, each with its own ceremonial setting. To follow the routines of what is in some particular milieu the normal day is in and of itself *prima facie* to act intelligibly; to depart from them requires some specific reason for so doing in order to be intelligible. Such reasons may be of several different kinds, but there is one species of such reasons attention to which is especially rewarding. They are the reasons offered, attended to and acted upon by those engaged in practices, in a sense of that word which I have defined elsewhere (*After Virtue*, South Bend, 1981, p. 175). Practices thus defined include such systematic types of human activity as archiecture and farming, baseball and portrait painting, physics and fishing. What matters about these types of activity from the standpoint of this present enquiry is that they are all types of activity within which individuals not only learn how to pursue certain specific kinds of good, but also how to transform earlier less adequate conceptions of those goods through rational criticism into more adequate conceptions. Thus activity within practices comes to be informed by both theoretical and practical reasoning. And understanding the relationship of such reasoning to the social contexts provided by practices is, so I shall argue, necessary if we are to understand the relationship of reasons for action to actions in a way that will throw fundamental light on the nature of intelligible action. But a worthwhile preliminary is to notice some additional points about intelligibility which have already emerged.

The first is that we exhibit the intelligibility of some action or type of action otherwise and hitherto unintelligible by producing what is recognizably a good reason for so acting. This suggests – no more than suggests so far – that the limits of intelligibility are the limits of the stock of what are recognizable as good reasons that we can produce or invent. The latter term is important for the stock of what are recognizable as good reasons can always be added to. But wherever at any given time for any particular group those limits lie, there perhaps lie the limits of intelligibility. It does not follow that to act intelligibly is always to act for a reason. Characteristically individuals do not have reasons for doing that which is part of the prescribed, scheduled routine of the normal day of their particular social group. Indeed the class of actions which comprise the normal day – and the normal month and the normal year – of some particular social group just *is* the class of actions which, when performed by members of that group, do not need to be supplied with reasons in order to be intelligible. The patterns of the normal day, month and year vary of course from culture to culture in their degree of rigidity, in how much and how little variation they allow to individual initiative. But they may extend to cover a good deal of informal and occasional activity. So going for a walk from time to time or on a spontaneous shopping expedition or taking a day off work may in certain settings be intelligible just as part of organized, routinised social reality.

It is not unimportant that everyday life of this kind embodies in its forms and patterns a great many of the shared beliefs of the particular culture and that these beliefs will characteristically be part of a set in terms of which reasoned justifications for the overall patterns of the normal day, year and month may be offered. But to say this is not to say that the provision of such justifications or even the ability to provide them is required for the actions of those who inhabit that particular social setting to be intelligible. There is indeed in almost every society, perhaps in every society, a general assumption to the effect that 'It is reasonable to live as we do', but this assumption, far from providing individuals with reasons for acting out their daily, monthly and annual routine patterns, makes the having of such reasons redundant, at least with respect to intelligibility.

Not that the occurrence of this kind of non-reason-giving, non-reason-guided, but patterned activity is irrelevant to the character of reason-giving; it is indeed the background presupposed and required by a good deal of it. For to act from a reason is to be caused by that reason not just to perform some particular action but to perform that action rather than what one would have done had that reason not been causally effective. That is, to act from a reason is to be diverted from the activity in which one was engaged up till that point. To cite a reason in explanation of an action thus always requires at least implicit reference to what the agent would have done in its absence, to a background of ongoing activity; and this background is often provided by the routines and schedules of the normal day, month or year, the *status quo ante* for departing from which we have to find a good reason. Why is it then that the reasons offered, attended to and acted upon by those engaged in practices have a peculiar importance in this respect?

It is because it is primarily within the context of practices that *good reasons* have to be sharply discriminated from other types of cause of action. It is central to initiation and education into practices that we have to learn both what counts as a good reason for acting in one way rather than another and how to be guided by good reasons. To be good at whatever it is – architecture, say, or farming or geometry or chess – is to be guided towards the recognized goods of the particular practice by dispositions informed by right reasoning, reasoning, that is, by the adducing of reasons that would be acknowledged as good reasons – although not necessarily as conclusive reasons – for acting in one way rather than another by *anyone* both educated into and acknowledging an allegiance to the goods of that particular practice. To learn to be good at a practice therefore involves learning how to discriminate between situations in which one is and situations in which one is not moved to action by *having a good reason*; and this is first learned by others giving one good reasons and criticising one if one fails to act upon them. What others do to and for one one learns how to do to and for oneself; and causal relations between events in the external social world are thereby replicated in causal relations between events in the internal mental world.

It is of course the case that we do encounter examples of causally effective practical reasoning in areas of social life which lie outside the context of practices, but in order to be able to learn from such examples we should have to be able to understand how they exemplify *general* forms of relationship between the relevant types of event and state; and the more closely that some area of human experience approaches the condition of a practice, the more likely it is that such understanding will be available. The generalisations which are crucial to the acquisition and exercise of practical rationality are of course rarely, if ever, universal, law-like generalisations; they are characteristically and for the most part "characteristically and for the most part . . ." generalisations. It is these that underpin the causal knowledge that we have to exercise in understanding the connections between *particular reasons for action* and the *particular actions* for which they are the reasons in characterising the actions of others, in recognizing what others are imputing to us and in our own reflective deliberations and explanations concerning what we are going to do and what we have done. The possession of such causal knowledge does not of course necessarily involve an ability to articulate those generalisations presupposed by particular causal claims. It does necessarily involve inner reflection and action informed by a not necessarily articulated or – by that particular agent – perhaps even articulable belief in the relevant true generalisations. How is such justifiable belief to be acquired?

It can only be acquired through training, through a variety of systematic interventions and comparisons of the outcomes of various interventions and the testing of hypotheses about the relationship of our interventions to such outcomes that we ever acquire anything worth calling knowledge about what precisely causes what. And in order that our interventions and comparisons be adequately systematic and that our testing of hypotheses be adequately stringent we need publicly applicable and checkable standards, such that one person can appeal to another's judgment. And the capacity for judgment on the performances of others – a capacity inseparable from the capacity for those judgments on oneself which constitute practical self-knowledge – is also something that has to be learned both from and through our participation in transactions with those others.

The social contexts in which the acquisitions of causal beliefs occur are of many kinds and there are some in which the relatively haphazard nature of social encounters and outcomes makes it seem unlikely that reliable causal beliefs, that anything worth calling causal knowledge, will be engendered in the participants. By contrast practices – and only practices – provide precisely the knd of social experience which affords the possibility of acquiring genuine causal knowledge and more especially that species of such knowledge whose subject-matter is the relationship of reasons to actions. For it is in the context of practices that we learn the systematic connection between our learning or perceiving that such and such is the case and our thereby having a good reason for doing so and so in this or that particular type of situation, and between the characteristics both of that situation and of ourselves and our success or failure in producing the action which we have a good reason to perform. We learn, that is to say, how both features of our environments and our own skills, attitudes and dispositions may obstruct or facilitate our doing that which we have a good reason to do. In the course of this type of learning within practices we have to make three distinct, but related transitions.

When we are first initiated into practices of any complexity, characteristically we do not understand a good deal of what we are required to do. We do what we are told to do perhaps under compulsion, perhaps to please others, perhaps, if we are young, to be doing the same as our peers are doing. But somehow or other we move from *doing X because A tells us to do X* to *understanding why in terms of the goals of this practice A has a good reason for asserting that X is what should be done* to *being able to judge for ourselves whether in terms of the goals of this practice we have a good reason to do X*. And through systematic correction we learn to evaluate reasons for action.

A second kind of transition is involved in learning from the experience of a certain kind of failure within a practice. We had judged, let us suppose – rightly – that there were good reasons for bringing about some particular state of affairs. We acted on these reasons in the best way that we knew how to act in accordance with what those reasons for trying to bring about that state of affairs had seemed to instruct us to do. But our actions were ineffective. The desired state of affairs was not the outcome. We are thereby forced to reflect on what it is that good reasons for bringing about that kind of state of affairs are good reasons for actually doing. We begin to understand better how to specify better the type of action that is to follow from a set of action-engendering, because reason-providing, beliefs. And this kind of improvement of the understanding is for the most part a matter of learning what constraints different kinds of social and natural context impose upon effectively embodying our reasons for action in actions.

Both these types of transition involve a third, that in which we move towards the acquisition and comprehension of an explicitly formulated body of generalisations which enable us both to articulate and to improve the stock of causal understanding that we have acquired. These generalisations are of two kinds: the constraints imposed by non-human nature exemplify the universal law-like generalisations with which the enquiries of the natural sciences furnish us; and the constraints imposed by the actions of others exemplify the same kind of "Characteristically and for the most part . . ." generalisations which guide causally effective rational agency.

The characterisation of the causation of human actions, insofar as they are directed by reasons for action, which emerges from this account, has one centrally important feature. We have noticed that when such events as *the giving of a good reason*, *the adducing of a good reason* or *reflection upon a good reason* and such states as *the having of a good reason* or *being disposed to be moved by a particular good reason* function as causes of actions, they may do so in either of two ways. What moves me to act on a particular occasion may indeed be *the giving of a good reason* or *the having of a good reason*, but the aspect of that event or that state which is causally effective in producing some particular action may be only accidentally related to that event or that state *qua* its being the giving or having of *a good reason for action*. What moves me to action when I have been given a good reason to act may instead be the tone of voice or the social prestige of the person who uttered the sentences which conveyed the reason to me; what moves me to action when I possess a good reason to act may in a similar way be some emotional residue, its origin long forgotten or never acknowledged which is accidentally associated with my having that particular reason to act. And in either of these cases what causes me to act is not the good reason *qua* good reason, but the good reason *qua* its mode of giving or of being possessed.

Genuinely reason-guided action then has to be understood as action causally produced by good reasons *qua* good reasons. And it follows from this that we have to reject Donald Davidson's account ('Actions, Reasons, and Causes' in *Essays on Actions and Events*, Oxford, 1980) of the relationship that holds between an agent's reasons for doing what he did and what that agent did, when the reasons provide a genuine explanation of the action. Davidson calls such explanations 'rationalizations' and he recognized that reasons for action and the actions which are produced by those reasons cannot genuinely be linked together by universal law-like generalisations, if we require of those generalisations that they characterise the causes in question as reasons and the effects in questions as actions. But Davidson is sufficiently a follower of Hume with respect to the nature of causality that he also believes that, whenever one event is the cause of another, it must be the case that *some* universal law-like generalisation specifies the relationship between the two events. His solution to the problem that he thereby sets himself is to hold that singular causal claims can be defended without invoking any law-like generalisation, but that, when such a causal claim is true, some such generalisation must link the event that is the cause to the event that is the effect, although not necessarily one which employs the concepts used in making the singular causal claim. "The laws whose existence is required if reasons are causes of actions do not, we may be sure, deal in the concepts in which rationalizations must deal. If the causes of a class of events (actions) fall in a certain class (reasons) and there is a law to back each singular causal statement, it does not follow that there is any law connecting events classified as reasons with events classified as actions – the classifications may even be neurological, chemical, or physical." (*Actions and Events*, p. 17).

Yet, if this were so, no reason would be causally effective *qua* reason in producing action and no action would be the effect of reason-adducing and reason-possessing *qua* the adducing or having of good reasons for action. Reasons would be causes only insofar as the giving, adducing or having of them was identical with some event that was neurological, chemical, physical or whatever, and actions would be effects only insofar as they were similarly identical. But if this were the case, there would be no such thing as learning to discriminate between being moved to do something by a reason *qua* reason and being moved by some other aspect of the giving, adducing or having of reasons and no such thing as learning how to act on the basis of reasons *qua* reasons rather than because of some other aspect. And learning thus to discriminate and thus to act is a central feature of the education necessary for the achievement of excellence within practices.

What Davidson's neoHumean account of reasons as causes has to exclude from view is the distinctive role of "Characteristically and for the most part . . ." generalisations not only both in guiding and in explaining actions, but also in thereby specifying the causal linkages which partially constitute the inner reflective life of the reason-guided person and which bestow on that life part of its essential coherence. By insisting that only universal law-like generalisations, of a kind which embody the canonical form of the statements at the core of the theoretical structures of physics and chemistry, can specify the causal linkage between reasons and actions, Davidson obscures the distinctiveness of the causality of reasons and reasoning in the human world. And he also thereby obscures the crucial link between the inner life of the mind and the social life of practices.

It is no part of my thesis that actions, understood and classified as actions, *cannot* have the kinds of causes whose relationship to their effects can only be specified by the formulation of some universal law-like generalisation or generalisations; it is no part of my thesis that actions *cannot* be that type of cause. What is true or false in claims about that kind of causal connection has to be discovered by empirical enquiry. But it is central to my thesis that such causality is not what distinguishes actions *qua* actions and that the type of causality which does distinguish them, the causality of reasons and of practical reasoning can only play the part that it does in that part of the life of the mind which consists of private, undisclosed reflections, insofar as that life is modelled upon and incorporates into itself what it has learnt from the type of education which only successful initiation into practices provides. Practical reasoning just *is* the kind of reasoning be learnt in practices.

To say this is to recognize that we are only able to know ourselves from within as rational agents because and insofar as we can be known by others as rational agents in virtue of our participation with them in practices. The rational mind is essentially social. But if this is so, then it ought to be the case that insofar as mind is not social it also will tend to fail in rationality, that in those human activities and situations where training in the kind of reasoning and the kind of reason-guiding that is characteristic of practices is absent, the ability to recognize connections between good reasons and actions will tend to be at a low level, at a very much lower level than the ability that is expressed in achievement within practices. The empirical psychological research which has a bearing on this issue seems to confirm this. One large group of studies show that individuals are to a remarkable degree, when guided by reasons, guided not by good reasons *qua* good reasons but by reasons *qua* utterances of prestigious figures or members of peer groups (for both a summary and a discussion of these studies see 'Social Psychology's Rational Man' by R.P. Abelson in *Rationality and the Social Sciences*, edited by S.I. Benn and G.W. Mortimore, London, Henley and Boston, 1976). Even more striking are the studies which show individuals as unable to report correctly what their reasons were for doing what they did (see 'Telling More Than We Can Know: Verbal Reports on Mental Processes' by Richard E. Nisbett and Timothy DeCamp Wilson, *Psychological Review*, Vol. 84, No. 3, May 1977: 231–257). What Nisbett and Wilson show, it seems conclusively, is that introspective consciousness does not of itself afford very much self-knowledge concerning the genesis of action.

Since none of the studies discussed by either Abelson or by Nisbett and Wilson used as their subjects individuals engaged in or developing the abilities acquired in practices, this does at least suggest that the kind of self-knowledge required for rational intentional action is to be had only in and through practices. Part of the importance of this conclusion, if it is indeed warranted, is that it enables us to understand further the close connection between and interrelationship of rationality and intelligibility. Consider next in this regard one obvious objection to my thesis.

III.

When someone adduces a reason for acting in one way rather than another, that individual is, so it is often supposed, necessarily giving expression to one or more of his or her desires. On this view the paradigmatic example of reason-giving is saying "I want (*or* I wanted) such and such." But such reason-giving is not, so it seems, in the least dependant on education into a practice. For among the wants expressed in our reason-giving are many which we share with young children. In saying "I want" in giving a reason I may well be expressing some universally felt need or lack, such as thirst, or some inner inclination towards some immediate pleasure or away from pain. Hence we tend to suppose that "I want this, so I shall do that" must express an unproblematic piece of practical reasoning and are apt to be disconcerted when it is claimed, by for example, G.E.M. Anscombe (*Intention*, Oxford, 1957. 64–5) that it does not express a piece of practical reasoning at all. My own view is rather different from Miss Anscombe's; for I take it that there are different types of practical reasoning and that *sometimes* "I want this, so I shall do that" can express one type of such reasoning (see 'How Moral Agents Became Ghosts', *Synthèse*, Vol. 53, No. 2, November 1982: 295–312). But there are contexts, and characteristically the contexts provided by practices, where it cannot so function.

Consider the following example. I am being shown by my fellow-scientists the first and so far the only specimens of a new hybrid fruit developed because of its possible ease of cultivation and food-value for peoples in some particularly barren and starvation-prone part of the world. I snatch them from the hands of the scientist who is preparing to analyse them and gobble them down. When asked "Why on earth did you do that?" I reply "I just felt hungry. I like fruit". It is important to notice that this answer renders my behavior more rather than less unintelligible than it was before. For in the context of this kind of scientific practice eating something just because one felt hungry is not an intelligible way of behaving. The practice imposes norms which constrain and limit the expression of immediate desire in the institutionalised social settings informed by the practice. Of course, if a young child had strayed into such a setting and similarly eaten a piece of fruit, not knowing what it was doing, the child's action would be perfectly intelligible. But the scientist whose understanding of what is going on is provided by the relevant forms of practice lacks any good reason for so behaving and the report of the felt lack of food, the felt hunger, even if true, does nothing to supply such a reason. That is to say, the citing of a want as such, independently of context, is *never* the giving of a good reason for an action.

To this it may be retorted that nothing so general is warrented by the example. There are obvious types of case where "I just felt hungry. I like fruit" can be a piece of reason-giving which provides all that is needed – if indeed anything is needed – to make my action intelligible. I am walking in my own orchard and take a ripe plum and eat it: "I just felt hungry. I like fruit". But now it is crucial that more has been supplied than the citing of a desire. What has been supplied is a social context and that context is one which makes it intelligible to consider acting or to act on such a desire, one which makes of the citing of such a desire a good reason. There are not contextless good reasons. And this dependence of a thought on social context for achieving the status of a good reason for performing some particular action is no less when the reason is one that I recite *only* to myself, only thinking and not saying it. Thus *having a good reason* is not in the ordinary stock sense of 'mental' (where 'mental' contrasts with 'physical' and every human property is either one or the other) a mental event or state at all. It is rather a social status which certain 'mental' events or states enjoy in virtue of their owners' particular place in some particular context.

The routines of everyday life and practices are not the only types of relevant context. For an action may well be *prima facie* intelligible by the standards of some practice and yet be unintelligible as done in these circumstances by that particular person with his or her particular history. An action can be unintelligible not on account of its immediate social context, but because of the agent's antecedent actions or experiences. It can be intelligible or unintelligible as part of the agent's personal history, as a sequence or part of a sequence in that agent's narrative.

Consider an example: someone buying groceries at a local store, as he has done for years, recognizes in the new owner someone who was a guard in the concentration camp where his wife was tortured to death. He goes on shopping there as part of his weekly routine, exchanges the same amiable casual conversation that he does with other store-owners and in no way ever treats this man differently from the way he would anyone encountered in this way. His actions, if they are indeed as I have characterised them, and if there is no significant additional true characterisation of them available, are *prima facie* unintelligible, given the standards of intelligibility imposed by contemporary narrative forms.

There is an important way in which the intelligibility conferred upon an action by its place in a particular agent's narrative is more fundamental than that conferred by its relationship to practices. For that an action is what a particular form of practice prescribes in this type of situation may satisfy a necessary condition for intelligibility and in many contexts may be sufficient to warrant a judgment of *prima facie* intelligibility, but by itself is never a sufficient condition of intelligibility as such. The question can always arise: why or how can that particular person with those particular narrative antecedents be intelligibly engaged in that type of activity *now* and *here*? Yet it is also true that the intelligibility conferred by practices has its own fundamental character. For without the kind of learning that is possible only through practices adult human beings do not come to have the lives of practical reasoners and thus the lives which are enacted narratives. The enacted narrative of a child's life is one in which the child itself as coauthor – and none of us are ever more than coauthors of our own lives – plays a relatively minor role. It is as and because the child ceases to voice only those wants which are felt immediately and comes to understand certain *goods* as its own goods (the account of goods presupposed here is that given in *After Virtue*) that its narrative takes on a form of its own. And this transition takes place as and after the child is initiated into and educated into practices, and learns to understand itself as progressing towards or failing to progress towards certain specific goods, among which is the good of understanding what a good is and perhaps what *the* good is. The application of the concept of intelligibility thus presupposes the application of the concept of a good, and a good reason is most adequately specified as a reason for doing something which will be or achieve some good.

That this is so gives us a good reason for rejecting an account of the relationship of the various subdisciplines within philosophy to each other that has been enormously influential (one version of it will be found in Michael Dummett's *Frege: Philosophy of Language* (London, 1973. pp. 666–672). On this account logic and the theory of meaning provide philosophy with its metropolitan centre, the philosophies of mind and action are to be found in the suburbs and ethics and politics are provincial, rural activities. So Dummett follows Miss Anscombe in speaking of "The dependence of ethics on philosophy of mind" (p. 666). But in fact this dependence is mutual and, if ethics was indeed sadly mutilated by its postKantian separation frommthe philosophy of mind, the philosophy of mind has been correspondingly deformed by its deprivation of all those concepts which find their place in the theory of goods and of the good and among these the concept of intelligibility is central. It is as important and as often necessary to correct errors in the philosophy of mind by considerations drawn from ethics as *vice versa*.

It is now possible to summarise those claims about intelligibility which are either explicit or implicit in the argument so far. Actions can be intelligible or unintelligible in at least three different ways: in the ways that they conform to or in the ways that they depart from the standards provided by the structures of everyday routines, in the ways that they conform to or depart from the standards of practices and in the ways that they conform to or depart from the standards embodied in narratives; but their intelligibility or unintelligibility does not arise from the simple facts of conformity or departure, but from the reasons which the agent has (or fails to have) for such conformity or departure. It is in virtue of the kind of reason for which an action is performed that it is intelligible or unintelligible. Hence the limits to intelligibility are set by the stock of possible – possible in that time, place and culture for that kind of agent – good reasons. It is thus not actions as such, but actions plus the reasons that the agent has for performing them that are intelligible or unintelligible. But to say this is misleading unless it is also understood that intelligible actions are the paradigmatic and exemplary cases of actions. An intelligible action is not an action (in the fullest sense) with something else added: intelligibility. It is rather than an unintelligible action is defective as an action.

What in any particular time, place and culture limits the stock of good reasons? The answer must be: the contemporary condition of practices and the contemporary availability of narrative forms. An appeal to basic physical wants and needs in an appropriate context may of course provide a good reason for action in *any* cultural order, although as I have already noticed there will always be some contexts which render such an appeal unintelligible. But the extension of the giving of good reasons beyond such wants and needs, either for departing from the routines of everyday life or for doing something within a particular practice or for carrying further the narrative of one's own life, depends upon the present condition of that practice and the presently available forms of lived narrative. Just as what is intelligible within the context of a given practice may be unintelligible (or else bear quite a different meaning) outside it, so what is intelligible at one stage in the historical development of a practice may be unintelligible (or else bear quite a different meaning) at another. And as with practices, so is it also with narrative forms. But this is not to say that the agent is restricted to the stock of reasons and actions employed within either framework up to that point, any more than a poet is restricted to the stock of idioms and forms employed up to the point at which he writes. To utter intelligibly as a poet, the poet must be uttering what is intelligible as poetry in the light afforded by the practice of poetry so far. Had William Dunbar or any other Scottish medieval poet uttered the sentences "My nerves are bad tonight. Yes, bad. Stay with me" he could not have recognized himself, nor would any of his contemporaries have recognized him, as practicing the art of the poet. But what Dunbar could do intelligibly and did was to extend the possibilities of poetry beyond those afforded by Chaucer and other predecessors. For the standards which define the practice of poetry do not define a closed domain, but rather one that can be indefinitely extended, albeit only stage by stage. And what is true of poetry is true of all other practices.

One postscript to the argument: it is worth remarking at this point that it was intelligibility and untelligibility in just this sense that Freud was concerned with in his discussion of neurotic symptoms. Freud's claims about the causal explanation of the occurrence of such symptoms depend upon a prior thesis about the redescription of the symptoms so that they become intelligible as actions. Prior to such a redescription neither the agent nor anyone else knows what the agent is doing. Thus Freud in identifying the neurotic symptom as an unintelligible action distinguishes neurotic symptoms from those parts of human behavior – twitches, sneezes, reflex responses – which are not actions at all. It is not clear that Freud himself clearly recognized this; his view that in the future psychoanalytic description and explanation might be *replaced* by purely neurophysiological or biochemical descriptions and explanations suggests that he may have not recognized the distinctive, irreducible place of the concept of intelligibility in classifying the phenomena for which causal explanations are to be given. Nonetheless Freud's own account of the neurotic symptom is a paradigmatic example of the use of the concept of *prima facie* unintelligibility.

It ought by now to be clear that, although one person can succeed in finding an action intelligible where another fails, perhaps by reason of greater or more accurate knowledge of the relevant context, actions just *are* or *are not* intelligible (this is not to rule out borderline cases). And among those doings which are or are not unintelligible are my acts of inner reflection, which like all my other acts can simply be unintelligible and therefore, if I myself have understood them correctly, be unintelligible to me, just as my neurotic symptoms will, in a state of preFreudian understanding, be unintelligible to me. Since the standards of intelligibility are the standards of the public world, of the realm of practices and of enacted narrative, it follows that insofar as my inner life is intelligible to me, it has been constructed and enacted in accordance with those standards. The boundary between the mental and the physical is not the same as the boundary between what belongs to the public and what to the private world.

IV

The general form of the thesis which I have defended is that genuinely human consciousness is such that our abilities to reflect upon and thus to be knowledgeably conscious of our inner lives – insofar as those inner lives are a matter of intending, doing, having reasons for action and believing – depends upon our abilities to interact socially with others within practices and to understand what they impute to us. We learn to recognize what is true about ourselves in this area and we learn to describe outselves only as others recognize and describe us and we learn to recognize and describe those others precisely as beings capable of recognizing and describing us in the way that they do. Self-consciousness is thus in large and crucial part secondary to consciousness of others as beings who are conscious of us; what I have called genuine human consciousness embodies a presupposition that the other minds problem has already been solved, or rather does not arise. There is thus a sharp contrast between genuinely human consciousness and what may be called Cartesian consciousness.

It is of the essence of a Cartesian consciousness that it is consciousness of itself unmediated by any other. Because a genuinely human consciousness knows itself in large and crucial part primarilty as a consciousness scrutinised by the consciousness of another, it can never know itself except as embodied. Embodiment is not a contingent or accidental property of consciousness, as it is according to both Platonists and Cartesians. John W. Cook's Wittgensteinian arguments reinforce this conclusion ('Human Beings' in *Studies in the Philosophy of Wittgenstein*, ed. P. Winch, London, 1969) and do indeed suggest that it might be extended to other contents of consciousness by learning from Wittgenstein's remark, quoted by Cook, that "Only of a living human being and what resembles (behaves like) a living human being can one say: it has sensations . . .." If Wittgenstein and Cook are right, then in one crucial way all the items of a genuinely human consciousness have to be characterised in a way that presupposes the possibility of their being ascribed to one by others, just because one's own avowals are secondary, derive their character from such ascriptions. So a general distinction would emerge between those items which are to be found in a genuinely human consciousness – after-images for example, which one has to learn to characterize and about the properties of which one may make mistakes that others can correct – and those fictitious items which furnish the contents of Cartesian consciousness – ideas, for example, if the furnishings are seventeenth century French, sense-data if they are twentieth-century Cambridge. It is this distinction between these two kinds of consciousness that makes it possible to return fruitfully to my initial question.

Why ought we to deny that my computer plays chess with me in the same sense that my son plays chess with me? 'Chess' is the name of a practice and to move a piece on a board is to perform an action which may be intelligible or unintelligible. Notice that a correct move can be unintelligible, that even the best move to make can be an unintelligible move for this player to make now: I am trying to amuse a sick child by playing chess with him. If I contrive to checkmate him immediately, given the reasons that brought me to his bedside, my action in doing so is *prima facie* unintelligible. Similarly an incorrect move can be an intelligible action: I am making a prearranged signal to an associate in the watching crowd. Conformity to the rules of chess is of course what makes my actions instances of playing chess, but it is not what makes them actions. They are actions only because and insofar as they possess or lack intelligibility. But the chess-moves made by my computer are neither intelligible nor unintelligible even although they always conform to the rules of chess. Hence we cannot ascribe the activity of playing chess – playing chess as a series of actions – to the computer, no matter what degree of skill it exhibits.

Notice that my thesis is not about all possible machines. If by a machine we mean only something constructed in a certain type of way to certain types of standard out of certain types of material, it may well be the case that one day something that is a machine thus understood will indeed perform actions. It may indeed already have happened; for on a certain type of theological view we ourselves just are such machines. But the machines that *we* make *now* cannot act and the whole progress of their development has not been in the direction of action.

Intelligibility depends, so I have argued, upon social context and beings capable of intelligibility in their actions must have the capacity to participate in social relationships and more specifically in those forms of social relationships which inform practices and enacted narratives. Machines, as we know them, have no trace of such capacity. They exercise numerous degrees and kinds of skill, but they do not acquire goals in the way that those initiated into practices acquire goals and they do not participate in practices in the way that those educated into practices do. I say 'do not' and not 'cannot'. For the problem is not that there is some demonstrable conceptual truth to the effect that intelligible action cannot be predicated of machines. It is rather that machines, as we now know them, lack the relevant kind of social history and the relevant kind of social relationships and that the traits which they do possess canot therefore be characterised against the background of such a history and such relationships, as is the case with distinctively human traits. Machines neither have nor lack good or bad manners, unquestioning or questionable loyalties, virtuous or vicious habits. They do not acknowledge or pursue goods. Thus it is not that we deny them any moral states *because* we do not ascribe a genuinely human consciousness to them. It is rather *vice versa* that we do not ascribe such a consciousness to them because they lack certain morally relevant properties.

Thus no degree of competence in the manifest exercise of any skill by itself provides sufficient reason to ascribe to any being the intentionality of a genuinely human consciousness; and it follows that Alan Turing's test ('Computing Machinery and Intelligence', *Mind*, 1950; Hilary Putnam has recently summarised a version of it elegantly: "let someone carry on a conversation with the computer and a conversation with a person whom he does not know. If he cannot tell which is the computer and which the human being, then – assume the test to be repeated a sufficient number of times with different interlocutors – the computer is conscious'' (*Reason, Truth and History* pp. 8–9) is never decisive in deciding whether something does or does not possess a genuinely human consciousness. For even a high degree of linguistic skill is only a high degree of a skill. But although to use Turing's test of the 'Imitation Game' to decide whether something does or does not possess a genuinely human consciousness is always a mistake, Turing may not have been entirely mistaken. For all that he proposed to ascribe on the basis of the 'Imitation Game' were intelligence and consciousness (in some far from fully specified sense). And I see no reason at all why what I have characterised as Cartesian consciousness and Cartesian intelligence should not belong to computers. The difference between human beings and machines – as we know them so far – is not that machines must be mindless; it is that they are incapable of action.

One reason why my computer cannot be said to be playing chess with me when I play with it is that, for all it knows, it may be playing with itself. It does of course possess a capacity for self-scrutiny: it is conscious of its own mental contents. It does not have after-images, but it does have ideas. A Cartesian consciousness is a solipsistic and ghostly consciousness and until it can formulate some valid version of the Ontological Argument it will be unable to solve the Other Minds problem. But ghosts of this kind are precisely the inhabitants that we can expect to find in machines with the capacities of the more advanced computers. Gilbert Ryle's famous characterisation of the Cartesian view of human beings embodied a very deep insight.

HOW TO INTERPRET ACTIONS

AMÉLIE OKSENBERG RORTY

The Crown Prince has inaugurated a new road by cutting the braided crimson ribbon. The road goes straight from the Royal Palace to the airport, a road that does not serve the City, a road built at great expense and coerced land appropriation. The Peasants, humbly dressed in grey rags and brown tatters, have been gathered together to watch: they've been rehearsed on when to clap, how to cheer. His part of the ceremony finished, the Crown Prince and his retinue return to the Royal Palace in their Daimlers and Bugattis. But the motley crowd of Peasants stay on, joined by the court Jester, flamboyant in red rags and purple tatters, who has come to join the celebration. Lifting her nose high in the air, she delivers herself of a set of rousing sneezes and coughs, runs up to wipe her nose on the edge of the ribbon, and then solemnly braids and ties it up again, carefully forming an elegant bow knot. At this, the Peasants begin to guffaw, to slap each other on the back, and to do elaborate summersaults. After about ten minutes of such hilarity, they sit down on the grass, and spend the rest of the afternoon picking the lice out of one another's scalps and braiding one another's hair.

I.

What has happened? How many events were there? What do we understand in it, of it? What remains to baffle us? And how do we begin to understand what initially baffled us? Significantly, it is events, indeed events in contexts, rather than isolated actions that we understand (or fail to understand). And so, too, we do not understand isolated strings of words, but conversations held by (types of) people in specific contexts. Of course we understand individual events – the newspapers are full of decontextualized reports of such events –, and we also understand utterances out of context. But when we interpret isolated sentences and events, we supply the standard contexts in which they normally occur. We fill in the story: and it is against the background of standard contexts that we fill in the details.

Margolis, J., Krausz, M. and Burian, R.M., (eds.), Rationality, Relativism and the Human Sciences.
 ISBN 90-247-3271-9.

After having worked in the field for a year, an anthropologist, watching the Inauguration of the Great Road, understands one level of the Jester's actions and intentions. Her gestures, her dress, her actions and intentions are intelligible, understood as variations on themes set by shared practices. The anthropologist also has a theory about the Peasants' lice-picking and hair-braiding: her thesis is that the hair motif – and its transformations in braids, in ribbons and in rags-and-tatters – provides the frame that defines events. She has an intricate and elegant account of the class-distinctive criteria for the individuation of events: Royal events, defined by the contrast between brilliant and dull colors, are segments of larger events which are themselves demarcated by Peasants' ritual grooming of hair. According to her, events are framed by motifs of fringes: the larger event begins with the cutting of the braided crimson ribbon and has its climax in the delousing and braiding of the Peasants' hair.

Are these events made intelligible – as MacIntyre has argued – by the implicit repertoire of good reasons shared by the Prince, the Peasants, the Jester and the anthropologist? And is all this intelligible to us because our actions and intentions are formed by interacting with these people in cooperative practices? What enables us to understand an action, and the intentions embedded within the event?

1. We begin by locating it in a web of culturally described presumptive classifications of action and events: the inauguration of a road . . . wiping one's nose . . . tying a ribbon . . . braiding hair. These give the rough, first, thin description and identification of the action type: the action is demarcated with a presumptive beginning, middle and end. It is located in a field of closely related and opposed actions. (The *inauguration of a road*, not its *military defense*; *wiping one's nose*, not *kissing a holy relic*. ("How are you?" is standardly interpreted as a conventional greeting, not as a request for symptoms one would give a doctor.))

The thin description of an action presupposes a background of a systematically connectd range of heterogeneous social practices: linguistic, pragmatic, and playful. These practices are not themselves atomically identified: they are individuated and identified by their roles in a cultural system. The thin description of actions carries a set of presumptions about the thin descriptions of the agent's intentions and expectations, about the significance of the action for her, and for those around her. The details of the event might well have different significance for the Crown Prince, the Jester and the Peasant. The various participants might have distinctive interpretations of the relations among practices; some groups might, and others might not, have a further understanding of the differences among their perspectives, a map of perspectives and their significance.

2. The thin surface description of the action – its typical classification and

identification – is thickened: horizontally, by a fuller description of the style and tonality of its performance. (The way the Jester gamboled to the ribbon, the elaborately mannered way she wiped her nose on the ribbon, the way the lice are crushed between the thumb and the forefinger.) This thicker description differentiates an individual action: it is distinguished from typical performances of such actions. The thicker horizontal description requires a thicker vertical classification: *satirically* gamboling to the ribbon; wiping one's nose *ironically*. As the action description gets thickened, the intentional aetiology is correspondingly thickened, well within culturally defined action-forms. As the description is more detailed, the hidden structure of the practices, their functions in the articulation of the culture emerges. As the action is located within the character structure of the agent, it is also located within the practices of the society.

But there are vertical layers in this thickening of the action. Briefly and schematically:

a) Species defined actions: Actions are located and interpreted within the context of cultural framings of constitutional structures. A sneeze is a sneeze: but it also has a complex theoretical description: it is the expression of an evil spirit; or the expulsion of phlegm in reaction to dust; or a genetically programmed reaction of the immune system to a specific chemical produced by the intrusion of certain sorts of particles within the respiratory system. What we call *dying* brings dramatic changes: but the details of those events – and whether they are classified as actions – are understood within the frame of a cultural system.

b) Variety defined actions: Actions are located and interpreted within the context of cultural framings of physical or psychological types or varieties: women give birth; men claim children by ritual *couvades*. (But whether a ritual *couvade* is a variety of parturition or an imitation of it, is fixed by cultural classification. (And this is quite compatible with there being a level of description in which there is a fact of the matter.)) Within the frame of a late mediaeval theory of temperaments, the smouldering rage of a smith of the Hephaestos type is distinguishable from the warlike fury of a soldier of a Martial temperament, and both differ from the biting anger of a melancholic Saturnine scholar, even when the thin descriptions of their actions would classify them together.

c) Culturally defined actions: Actions are interpreted by the cultural repertoire of narratives that locate the sources and consequences of events: they define actions by defining stories of agency set by characters, roles, occupations, age, gender. Thinly described, Marie Antoinette can think of herself as performing the actions of a milkmaid, but a thick description distinguishes the actions of a milkmaid from those of a Queen imitating a milkmaid. Thinly described, the Royal Hairdresser and the Peasants all are *braiding hair*; but a thick description distinghuishes the significance of Royal acts from that of the Peasants.

Differences in significance are standardly located within the repertoire of cultural narratives. Why did she do that? The details are filled in by subsuming them under set stories – the Bible (Ruth is her prototype), the latest movies (she's just seen a Katherine Hepburn movie). Locating actions by such narratives are interpretively enlightening for a number of reasons: they provide implicit sensory detail with which to envisage the sources and consequences of the action. Vividly imagining the narrative context of an action is one mode of understanding. But even more significantly, such narratives present self-fulfilling accounts. They explain actions because they did in fact also form them. Agents model their actions by standard cultural stories: in situations of uncertainty or choice, it is just these stories that guide action. Because agents improvise their actions by improvisations drawn from this stock of cultural narratives, the stories provide one basis for interpretation. When the interpreter locates the causally appropriate narrative – the narrative that constitutively forms the action – his interpretation is not arbitrary, not imposed.

d) Individually defined actions: Actions are located and interpreted within the frame of an individual's history. Of course such a history connects the species, variety and cultural factors: a person's actions can sometimes be explained by accidental narrative crossings of these factors. The agents' interpretation of these accidental events sometimes further individuates them, and a full understanding of the significance of the action may require understanding the interpretation that the individual gives to the accidental events that compose her history. Second level interpretations of individual agent's interpretations are, of course, difficult to formulate, let alone to verify. But often the association of ideas, the sequence of thoughts, the imagery of the agent's descriptions help to fill in the context. Some of these details are horizontal: they link the context of the action to a wider network of actions on the same level of description. Others are vertical: they further specify the aetiology of the action.[1]

3. Interpretations of actions rest on both formally invariant and culturally defined substantive conditions for rationality (MacIntyre's "good reasons"). Even the thick identification of actions also locates them within a set of rational derivations. An extension of the Principle of Charity guides interpretation: there is a presumption of 1) the preservation of truth; 2) the preservation of a system of valid inference; 3) a motivational structure which is systematically directed to (culturally and intentionally defined) thriving.

The Jester moved towards the ribbon because she could use it to wipe her nose, but she moved to *that* bit of cloth because it played a role in the Royal ritual. The specification of her actions embeds practical reasoning: ordered beliefs about the properties of objects, the effects of certain sorts of movements, how standard actions and words are ordinarily understood. Interpretations of actions include, and indeed, presuppose standard forms of such beliefs and inferences. The rich detail of the Jester's movements – her making a direct line for the ribbon, rather than her meandering about for half an hour before reaching it – follows from presumptions about what can be reconstructed in a system of practical reasoning. This is not to say that such actions are, in the strict post-Humean sense of the term, caused by such reasonings: the Jester need not have asked herself, "What is the best way to get to ribbon?" She need not even have asked herself "How can I best mock the outrageous absurdity of this road?" Yet her actions are formed and directed by the background presence of such reasons. They are implicit in her (socially formed) repertoire of beliefs and sensibly grounded actions. She learned what to do, how to go about doing it, and how to speak about what she does by taking part in family life, in the life of the community. What formed her directions and expectations also provides her understanding and interpretation of what others around her do. The conditions for the specification and determination of her own actions – their being nested in the system of her purposes, habits, her typical roles – are at one and the same time the conditions that enable her to understand and interpret others.

The anthropologist, observing the Jester and the Peasants, attempts to reconstruct what is implicit in the Jester's practical reasoning, the immanent structure of her thought and actions. The anthropologist manages to reconstruct the Jester's practical reasoning by coming to share the social practices of the community, by interacting with its members, by being infantilized by them and by (corrigible) presumptions about the focal and salient features around which cultural interpretations cluster. ((What we call) birth, puberty, sexuality, marriage, the definition of power, exchange and inheritance, certain forms of play, death are given form by virtually all cultures and subcultures.) The anthropologist begins with presumptive interpretations of these, interpretations which certainly initially have their focus shaped by the preoccupations of her own culture and personal history. But as she lives with the natives and interacts with them, she comes to share in their history, to partake in their common references. While some of her interpretations will no doubt have their proper place only in her own culture, they will also be formed by the give-and-take of her interaction with the folk.

4. Interpretations also depend on certain forms of improvisation, play and formal transmutations and transformations. Nothing mysterious is introduced by the notion of formal improvisations. The *formal* is detachable from its contexts, can be varied independently and serves as the model for distinctive instantiations or realizations. Such formal transformations, grammatical-spatial rotations – inversions, conversions, reversions, subversions,

obversions, abversions, puns, fugues, palindromes, turning subject into object, changing the tense, changing the power relations and order of dependence, reversing up and down, miniaturizations, blowups, and distorting ripple mirrors of all kinds – are as central to the formation of action (and therefore to its interpretation) as the grounding of practical reasons.[2]

We can understand small imaginative transformations, as forms of play. Transformations which occur a step at a time, an improvisation at a time, are made intelligible as permutations of the previous moves. They can arrive at actions which are radically different from their origins, and which are intelligible only by tracing the intervening transformations. Such improvisation, formal permutations, questions, and irony all have relatively formal rules, and these can vary culturally. Transformations lead to actions which cannot be reducibly interpreted as expressions of species formation, or of culture-preserving and reproducing practices. A Buster Keaton of the yo-yo can do Round the World elegantly, with verve and flash. Then as a matter of improvisatory style, he adds an extra loop, a crawling zigzag at the apogee. Or Charlie Chaplin wriggles his mustache and eyebrows, swings his cane, and suddenly turns his body into a pendulum. These are actions, they are intelligible as improvisations on standard modes of action; they are intentional deviations, not grounded by reasons to narratives or routines, not even to personal narratives. Interpreters can give subtle and precise accounts of the significance of these improvisations. Such arabesques are the raw material for the pattern-discerning work of literary critics, anthropologists and psychologists. Transformations may appear within the structure of rationality (as manifestly they do in grammatical construction and in the transformations traced by formal logic), but they are not grounded in, nor made intelligible by, a culture's stock of practice-based good reasons.

Such play forms Socratic irony; it gives sense to the nonsense of Lear's fool; it informs Meredith's claim that satire is the most powerful weapon against folly and injustice; it is the crux of Bergson's analysis of comedy, and Freud's account of the wit with which the unconscious manages to get itself heard. What Mondrian does visually, by transforming sunsets against trees into black lines containing red squares, what Picasso does in transforming a bicycle seat and handlebars into a bull's skull, what the punster does verbally, the Jester does with her gesture: all are intelligible. Even when the Jester has good reasons for making authority ridiculous, the detailed intentions through which and by which she mocks established practices sometimes just are elegant formal jokes.

Our reasons sometimes stand behind, and not within the high-wire antics of intelligibility. That this is so, is our breath of freedom from the dangers of charisma and power politics that standardly pervade the social practices of evaluating reasons. The practices surrounding the evaluations of good reasons are sometimes not legitimized by the reasons they purport to assure: and yet it is often the practices rather than the reasons that dominate. They are fortified and defended against attack by good reasons because they illicitly introduce the very criteria for the legitimation of reasons which are at issue.

These forms of play, and this play of forms, are also often reductively made intelligible within the context of social and culturally formed practices. As the ambiguities of serious moral reform can transform the contexts in which they occur, so can improvisatory joking break the very practices that frame it. Because sheer improvisation is intelligible, serious moral reformers and sheer improvistory clowns can make themselves intelligible by asking the undermining question, shift the burden of proof, declare the Emperor unclothed. Yet in a sense the past has been preserved through these transformations. While transformations can bring us to new forms of actions, the old forms are still discernible within their transmutations, and there is a power struggle between the reductive and the transformative interpretation. Indeterminacy of translation is often a function of a power struggle between two modes of interpretation, between older social practices and their formal transformations. Revolutionaries and reformers depend on the ambiguities of their actions and utterances speaking with the vulgar in order to bring them to understanding. The interpreter attempting to understand actions that fall both within and outside standard cultural practices recreates the steps of the transformation, by tracing the details of their formal permutations.

5) But sometimes interpretation also introduces theories that are not part of the native players' repertoire of good reasons, narratives, practices, formal transformations. An historian, an anthropologist, a psychologist can make actions and intentions intelligible in terms quite different from those used by the players. Braudel, Anna and Sigmund Freud, Marx, a couple of Geertzes can make the actions and intentions of distant people intelligible to us even when neither they nor we share the practices of these distant people, except on the most general level of description: eating, giving birth, dying. Canny interpreters can discern structures hidden to players, and sometimes alien interpreters can explain the self-deceptive characterisations of tenaciously powerful players' "good reasons." To be sure, what makes actions intelligible must have been causally efficacious in determining them. But since native players are not always the last or best authority on the causes that move them, the causal factors required to explain their actions can be articulated in theories that only use the Player's practices and reasons as identificatory starting points. Under such circumstances alien interpreters can have a clearer understanding of the intelligibility of the players' patterns of action than do the players themselves.

General scientific theories interpret the sequences of the events by analysing their stages and components. They define analytically independent variables which can in principle be reconstructed to project predictions. Hume put it most clearly: one form of understanding just consista in being able to connect phenomena in repeatable, predictible patterns. This form of understanding acquires stronger claims to objectivity when analyses, regroupings and reconstructions enable us to carry out practical intentions. A theory that enables us to fly from Omaha to the moon and back endows its primitive terms with significance: the analytic variables *gravity*, *force*, *matter* themselves become comprehensible, interpreted by what they have explained. To be sure, such explanatory theories are to be found primarily in the natural rather than in the human sciences. Nevertheless, interpreting and explaining actions presuppose and embed such physical and biological theories: after all actions are (also) physical events.

Though there are strong resemblances, explanatory theories differ from explanatory narratives: they introduce analytically distinct variables whose independence cannot be readily reproduced in nonexperimental situations, in ordinary experience. Narrative explanations are essentially temporal: an event is explained by its antecedents, and no explanatory variable is introduced that is not itself part of a temporally defined narrative sequence. An explanatory theory can introduce analytic variables that do not correspond to phenomenologically experienced events. There is, furthermore, a presumption of the Unity of Science: because explanatory theories are presumptively located within the frame of one complete and general theory, they are constrained and supported by their connection to related theories. But while explanatory narratives often feature the same characters, refer to each other and establish a continuity of action, there is no presumption that they are all part of One Great Narrative.

II

And it is not merely a matter of translating, understanding and interpreting actions from an alien culture. What does it mean to understand something familiar yet remarkable and strange in our own culture? One day our friend is amongst us, full of laughter and despair. The next, we learn that he is dead of a heart attack. We know that he smoked, took no exercise; that he ate rich, unhealthy foods; that he flamboyantly, defiantly neglected his health. Is there a sense in which his death is an action that requires interpretation? What would it be to understand that? These questions are not of course an invitation to mystical poses, reverential silence before the ineffable. We do in some way absorb these events. They form the substance of our lives, and we absorb them, act on them in such detailed ways that we know, more or less, what to do next. In culturally formed grief, we close the eyes of our friend. We read parables and novels, which somehow shape these events, place them in a story. We tell these stories again and again; we represent and re-experience them dramatically. Before we witnessed the death of a friend, we had some interpretation, some understanding of these events, if only that sort of understanding that is provided by a certain kind of familiarity. (Yet there is also a level of incomprehension, or at any rate, of silence.) Our general scientific theories supplemented our repertoire of narratives to prepare us: the connection between cholersterol diets, inherited genetic factors and heart attacks was established.

The images for interpretation – the images embedded within literary and epistemological theories – are surprisingly few, and surprisingly static. From Plato, we have the image of understanding as being-formed-by; Aristotle adds the images of becoming-identical-with and absorbing-or-grasping. Vico introduces the image of understanding as a form of making. Dewey improvises on Hegel to combine all three into an image of continuous interaction, an interaction which involves mutual formation. Such interaction changes both parties, and it is through these small steps of inter-change that what we call interpretation and understanding take place. Older images have been misleading: it is not as if the world is there, either affecting us, or not-quite-there, to be further determined by us. Both occur, as they do in the exchanges of genuine conversation. (This is not the Great Conversation among the Great Expounders of Great Systems, but rather the conversation that takes place among friends, who listen to one another, before they know what they want to say next. In contrast to an exchange of monologues, such exploratory interchange brings about understanding by producing mutual transformation.) The response of the other – whether it is the native Players, or Nature, or our neighbors – helps us understand the direction of our own setting forth, helps to make it what it is, to determine it further, by determining the next move at a time.

MacIntyre is right: the intelligibility of individual intentions rides on publicly shared meanings implicit in social practices. The social interactions that structured and formed the standard characteristic connections between intentions and actions also fixed the terms for understanding intentions and the actions. Besides being located in a taxonomic system of means to ends, and besides being located within scientific theories, actions can also sometimes be made intelligible by being located in a stock of culturally familiar narratives, or by being placed in the context of normal routines, expectations of what various types of people do in specific sorts of situations. Our understanding of these actions follows from having been ourselves formed by those practices: it is by sharing those practices, acquiring them in childhood, that we became who we were. So at least parts of our conversations with others, are constitutive of those inner conversations that form our own intentions and actions. But the system of cultural practices is (as certainly MacIntyre knows) not entirely determinate or fixed. If the cultural actions patterns that guide interpretation are set in the system of cultural practices, that system is itself continuously in dynamic – and sometimes contentious – formation. These are not the practices of a culture which has inherited a perfectly formed world, but rather the practices of a people forming their culture, who often find themselves significantly divided among themselves. To interpret and understand their actions, it is necessary to see them as also engaged in polemical conversations about the central features of their cultural practices. To understand the actions of the Jester and the Peasants, as well as those of the Crown Prince, it is necessary to interpret their misunderstandings, and struggles to define the character of their common practices. For this we need a range of quite heterogeneous skills: the task of interpretation requires all the help we can get. Attempting to reduce all of these to one set of focused heuristic strategies seems an unwise, ill-advised decision, on a par with trying to make a homestead farm flourish with the use of only one tool, or to reproduce a Pollock in black-and-white.

NOTES

1 In all this, we are still well within the terrain of interpretation described by MacIntyre. But he plays down the interpretive role of species or biological theories in order to stress the role of cultural practices and institutions that preserve the basic assumptions and directions of the culture.

2 Cf. Winnicott, Huizinga, Schiller on the liberating power of aesthetic play and improvisation.

MIND AS A SOCIAL FORMATION

ROM HARRÉ

The social construction of mind is the thesis that mind is a cultural product the genesis of which in individuals is closely bound up with the social and historical condtitions within which those individuals grow up and are constrained to act. Mead (1934), Vygotsky (1967) and Wittgenstein (1980) have each contributed refinements and clarifications to the thesis that the mentality of individuals is a product of their circumambient social condtitions. Can these various insights be brought together into a systematic account of mind sufficiently clear and simple to serve as the basis of research programmes in psychology? In this paper I shall try to show how, by bringing out the distinctive metaphysics implicit in the social construction point of view, and by setting out a clear alternative system of conceptual controls to the Cartesianism that animates most of recent psychology, the main projects of such a programme can be delineated.

A. **The metaphysics of conversation**

I take the metaphysics of a science to be the open set of background assumptions that controls both the theoretical and empirical practice of a community of scientists.

Recent 'experimental' psychology has been based on the same metaphysical scheme as that of the classical physical sciences. People are taken to be internally complex things, located in a physical space-time, the space-time of Euclid and Newton. Whether the internal complexity of people-things is describable wholly in terms of the concepts of the physical sciences or whether some residual mentalistic concepts are needed, is an issue contained within this general framework. One effect of the unexamined acceptance of this metaphysics is to direct research concentration on to individual internal causal mechanisms, be they physiological or cognitive, Humean or generative. Social constraints on thinking, feeling, intending and acting appear as addenda which would be eliminated by the methodology of individualistic experimentation and relegated to the status of *ceteris paribus* conditions.

Margolis, J., Krausz, M. and Burian, R.M., (eds.), Rationality, Relativism and the Human Sciences.
 ISBN 90-247-3271-9.

From the social constructionist point of view this metaphysical scheme is radically inappropriate (as is its Cartesian sibling, the subjectivism that underwrites phenomenology). An important manifestation of mindedness is speech, perhaps the most important of the human social practices in which mindedness appears. Let us begin with an analysis of a conversation among (not 'between') a community of speakers. By a 'conversation' I mean the playing out of a Wittgenstein language-game, that is the using of speech within and integral to some activity which is a recognisable practice within a form of life. For instance, the instructions, orders, questions, classifications, reprimands, etc., integral to the communal practice of building a house, christening a child, and so on. The animating thought is that – in many ways it does not matter *where* in Newtonian/Euclidean space-time, an insult, an order to fire, etc. is delivered, but *who* delivers it and to whom. It is the fact that I insult you rather than that the insult occurs in the dining room instead of the car that matters. Only certain people can bring about nuclear annihilation by speaking and their speech counts whether they are in Camp David or the White House. Plainly, indexicality is going to be a major clue to the metaphysical constraints I want to bring out.

Let us try out the idea that the people participating in a conversation are an array of locations for the speech-acts that are their conversational contributions and compare this with the Newtonian scheme.

The physical world is manifested in an array of locations, which are the places and moments of a spatial-temporal grid. These places (and moments) are picked out by indexicals (typified by 'here' and 'now') whose sense cannot be reduced to geographical and calendrical references, since they contain information, namely of how utterance is related to speaker, not conveyed referentially. The world is constituted by causal interactions between internally complex things variously located in the spatio-temporal array. The question of whether things can finally be eliminated in favour of colocations of causal interactions is open.

The social-psychological world is manifested in a grid of locations which are the array of people of a social order. People are picked out by indexicals (typified by 'I', 'you', etc.) whose sense cannot be reduced to proper name referents. (If I say 'Rom Harré insults you' our subsequent relationship will depend upon whether you know that I am Rom Harré.) The world is constituted by the speech-acts of the community in which a conversation comes into being. This 'world' can be extended to all interpersonal exchanges that involve displays which are intended to have social significance and which must be complemented by uptake amongst the other participants in order to be socially effective.

Assigning priority to this second scheme for the purposes of delineating the project of the human psychology has the immediate effect of suggesting that human interactions will, amongst other constraints, be conducted within the conventions of conversations. For instance, for certain sorts of conversations intelligibility, sincerity, turn-taking, etc., but for others different conventions will be in force, e.g. telling exemplary anecdotes, witty exchanges, shouting people down, etc.

I have developed the parallel between Newtonian ontology and a conversational community only with respect to space as the set of possible locations for things. The temporal aspects of speech-acts in an array of people-places are a topic in their own right, of some complexity which I shall not address in this paper; for instance, is there anything corresponding to absolute time? (See my *Personal Being*, ch. 2; Harré, 1983.)

The conversational 'world' is created by an interpersonal relational structure of speech-acts (and other symbolic displays) and their understandings. But in the 'stripped down' metaphysical cheme I am playing with, people as places are simple entities, without internal states. Intentions to speak or act with such and such social force must be treated as public Wittgensteinian first-person avowals, and beliefs as to the speaker's or actor's intentions as displays of appropriate complementary speech-acts or actions, and even perhaps public displays of understanding such as sage or sympathetic nodding. The pair 'intention-understanding' is semantically cemented in the conversational world, just as the pair 'cause-effect' is physically cemented as the operation of some generative mechanism in the physical world. (See Searle (1983) for the idea of an 'intentional structure'.)

The people-array, the set of locations that corresponds to space, differs in an important way from the Euclidean flat space of the Newtonian scheme. It is not isotropic, that is it has a structure by which one region differs from another. The people-array of a psychological space is structured by differential *rights*, *obligations* and *duties* to speak, and reciprocally rights, obligations and duties to listen. In most societies certain topics are reserved to particular persons. One cannot, without impropriety, reprimand someone else's child. In terms of the metaphor the space of conversational interchanges is 'curved'. The curvatures of physical space are associated with gravitational fields. The structured psychological 'space' is associated with moral orders since the structure of person-array follows from the existence of differential rights and obligations both to speak and to listen.

Empty places are silent people without the right to speak. But people may be silent for a variety of reasons. There are the aphasic, the shy and the sheepish. More importantly there are those who though they have the capacity, even the urge to speak, have no right to do so. A member of a church congregation is a location in psychological space, a possible place for the utterance of a speech-act, but not having the right to take part in the litany that place remains empty during the episode. The principle that children should be seen but not heard creates a few empty places in some person-arrays.

There are parallels, too, to the anistropy of contemporary conceptions of space. Newtonian-Euclidean physical space is everywhere the same and its properties are indifferent to translation and rotation of coordinate axes. But conversational space is like the curved space of general relativity. For example, if one's phone is being tapped speech acts travel from the speaker to the agents of MI5, who it is presumed have a right to listen, all in one direction. No complementary speech-act, say a warning, returns back up the wire. The 'space' slopes steeply towards the eavesdropper, like the distortions of space in the neighbourhood of a black hole.

The ontological independence of the conversation as a public and collective intentional structure is required by the possibility that there are structural properties of the conversation that are not reducible to the intentional states of the speakers. There is any amount of evidence for the existence of conversational properties that are not conjunctions of the properties of individual speech acts. A conversation may be produced which has cognitive properties of rationality, etc., independent of the intentions and cognitive operations of individual members. But more convincingly still is the work of Pearce and Cronen on the analysis and reconstruction of conversation between intimates which has shown that the stability of a relationship may depend on properties of the conversation unintended by either of the participants in preparing their contributions and mutually opaque to each of the speakers.

Even though everything that takes place in a conversation is and must be materially grounded, criteria of individuation and identity cannot be found in the material properties of speech as noises in the air, and the physiology of utterance and uptake. Furthermore, speech acts may differ, not only by their decontextualized lexical meanings, but also by the nature of their neighbourhood in a conversation, in particular by their relations to other speech acts. Speech acts form part of an interpersonal, public and collective intentional structure.

Taking for the moment the independence of the conversation as a working hypothesis, there are some methodological consequences to be noted. Considered as locations for speech-acts people are simple elements of the system. This is plainly an artificial assumption. But it permits the formulation of a range of questions whose possibility is concealed by the traditional metaphysics, in which people as things are to be treated as internally complex mechanisme whose public performances are the end-products of individual cognitive/affective states.

How much of that which we have taken to be researchable as cognitive attributes of individuals is to be found in properties of the conversation? Further elaboration of this point can be made by using Wittgenstein's distinction between public avowals and subjective descriptions. The former are typical of first person contributions to a conversation, the latter of third person remarks. Whereas 'I am in pain' is part of the expression of pain, and not a defeasible judgement based on evidence, 'he is in pain' is an empirical statement about his subjective state based upon evidence of, amongst other things, his public avowals. In this case Wittgenstein would have been the last person to deny that the speaker felt something unpleasant. But there are cases where a first person avowal is meaningful even when there is no subjective counterpart. For example, certain speech acts such as 'I am going to . . .' are conventionally taken to express intentions. They can become part of the public record, so to speak, and are binding on the speakers whether or not there was a consciously entertained intention as its subjective counterpart. Others may maintain the resolution of the speaker even if he or she has forgotten the occasion of the speech-act by such reproaches as 'Daddy (Mr President) you *said* you would . . .', by reference to some written or other public record of the utterance or writing of the speech-act in question.

Could there be a culture in which *all* matters that we take to be psychological, for instance rational (or irrational) thought, the labelling of feelings as emotions, etc. are carried out publicly and collectively through inter-personal conversational exchanges? Such people would be only physiologically complex. Their utterances would be thoughtless and spontaneous, but spoken in accordance with social conventions, though not consciously intended as such. I do not suppose that there is any such tribe, but it is a possibility within the conversationalist ontology and provides the necessary foil for the question of how psychologically complex with respect to what kinds of episodes and topics different categories of real people are. There may be enormous cultural diversity.

Philosophers have tended to treat memory as if the deep issue was the problematic grounds for claims for the verisimilitude of private recollections of events whose retreat into the past necessarily makes them unavailable for current checking. But the psychology of memory as a real human faculty cannot properly be confined to that sort of epistemology. The social constructionist point of view helps to draw attention to another aspect of the way recollections are certified as true records of past events. Memories are those reminiscences (conversational contributions prefaced by 'I remember (etc.) . . .') which have been authenticated by those who have memorial rights. Typically in a family the mother exercises memorial rights and reminiscences become recollections once they have her imprimature (cf. the work of M. Kreckel (1981) on performative speech in family circles). Again, we run across a matter of social structure and moral order (differential rights) in the 'real life' working of a phenomenon which has traditionally been studied by experimenting with the recollective capacities of isolated individuals. The point is not to deny that such capacities are of interest, but to emphasize that the language games of remembering cannot be explained by reference to those capacities alone.

These points have come up through exploring some of the consequences of taking individuals as simple locations in people-arrays at which speech-acts can occur. But people are psychologically complex. Our scheme must now be elaborated to make room for hypotheses about various ways that the psychology of individuals is, or could be, thought of as socially constructed.

B. A conceptual 'space' for psychological concepts

The conceptual schema upon which both subjectivist (phenomenalist) and objectivist (behaviourist) psychologies have been constructed is a linear 'space' based on a polar opposition between two 'realms', the 'inner-subjective' and the 'outer-objective'. These realms are asymmetric since each inner-subjective realm can be known only to the individual whose realm it is, while the outer-subjective is knowable by anyone. Thus we generate the Cartesian problem of 'other minds' in its simplest form. (For an interesting comparison of the projects of Marx and of Wittgenstein opposing that opposition, see D. Rubinstein (1981).)

To help define the projects of a social constructivist psychology I propose a three-dimensional 'space' based on the following oppositions.

1. A dimension of display, having public manifestation at one pole and private keeping to oneself at the other. For many psychological purposes it does not matter how something is concealed – behind one's face or behind one's hand (cf. Wittgenstein on locking up one's diary).
2. A dimension of realisation, having matters which are individually realised at one pole and those which are collectively realised at the other.
3. A dimension of action, representing the stance of the individual to the matter in question, whether as active agent or as patient. Diagramatically we have the following picture:

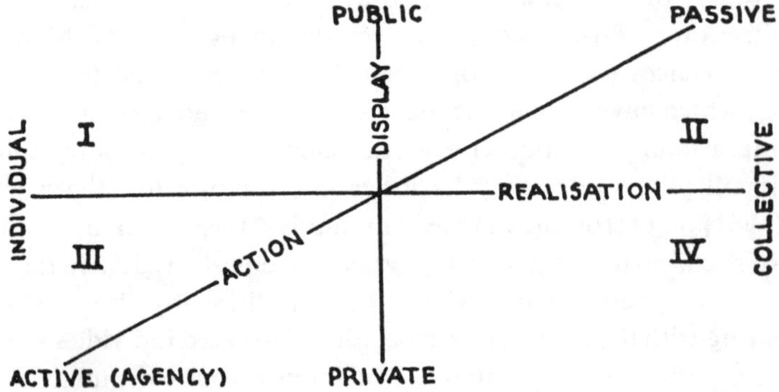

There are two uses to which such a scheme can be put.

(i) to control conceptual analysis and through it, research design. Knowledge though not a popular topic of research among psychologists, has an interesting duality when analyzed within this framework. Philosophers, working within a Cartesian metaphysics, have analyzed knowledge in terms of the individualistic concept of belief, coming up with such definitions as 'knowledge is true or certified belief'. In so far as there is a psychology of belief (cf. Dennett, 1979) the psychology of knowledge would be part of it. But the knowledge picked out by this concept is typically an individual person's store of certified beliefs, the sort of knowledge that is assessed by a written examination. However, a glance at the variety of language-games in which 'to know' appears, shows that there is a public-collective practice in which performatives of trust such as . . . I know . . .', meaning roughly 'You can take it from me . . .', etc. are used to create a shared corpus which can be used as a resource by actors in various tasks and contexts. Sociologists have shown that such a corpus is socially structured by a complex moral order of rights to assert and rights to display, which belong to different categories of persons differently. For example, there is lay and professional knowledge, common and esoteric knowledge, men's and women's knowledge (cf. Schutz and Luckmann, 1973).

(ii) This scheme can also be used to represent that process which I have called, following Mead and Coulter, the social construction of mind. As a thesis of psychological development it was already fully formulated by Lev Vygotsky (1962). According to the Vygotskyan point of view, human psychological development proceeds by the imposition of order and structure on inchoate thought and feeling. This order has its source in the public world of material and conversational practices. In Vygotsky's view the main instrument for the social construction of mind was speech, the acquisition of which shapes the inchoate flux of feeling and cognition into that we recognize as mind through acquisition of grammatical models and conventions of address. Wittgenstein seems to have had much the same insight; coupled with the insistence that one must always bear in mind that speech occurs in social/practical episodes, the so-called 'language games'. Mind is a personal appropriation from social, that is public-collective practices.

But what sort of practices would it be appropriate to pick as possible progenitors of the mentality of individuals? If this mentality is a secondary psychological structure from what features of the primary structure of public conversations is it appropriated? The Wittgensteinian notions of language-game and form of life are germane but both need spelling out if this conceptual scheme is to sustain any interesting research programme. By 'language-game' I take Wittgenstein to have meant any social practice (for instance building, counting, christening, and so on) in which speech plays a significant part. (We might be wise to widen this to include other semiotic systems than speech, such as gestures, road-signs, heraldic devices, etc.) By 'form of life', a notion central to but vague in Wittgenstein's thought, one can usefully mean roughly what anthropologists mean by a culture, encompassing every kind of language-game, from agricultural practices to beliefs in a cosmic realm, but within which, for reasons which will emerge, I wish to emphasize the presence of a moral order, particularly a certain distribution of rights and duties, and criteria for the moral evaluation of persons, their actions and their feelings.

What is it that language-games offer for appropriation by an individual as forms to shape generalized mentation into a personal mind? I can briefly touch only on four.

Public-collective conversation includes ways of speaking (and writing) that can serve as grammatical models for organising concepts of the mind. For instance, I believe a strong case can be made, both philosophically and anthropologically, for the view that the grammar of personal address in use in the practices of the public-collective realm, serves as a model for the organisation of individual thought. The typical grammatical forms by which we learn to transform avowals into self-ascriptions of psychological states contain the structures that reappear in the ordering of thought around a private inner self and so account for the structure of consciousness. The form of the latter should vary with the structure of the former if this thesis is correct.

C. **Four modes for the social construction of minds**

The public conversation includes a repertoire of grammatical models, a store of exemplary episodes, a set of conventions of discourse and an array of interpretative concepts. Each of these can play a role in the organisation of naturally engendered fluxes of subjective qualia into minds.

Case 1. Grammatical models

Imagine an infant in a social world. Among the linguistic features of that social world are grammatical models for the performance of certain kinds of speech-acts appropriate to the realization of certain intentions. There is also a moral order with rituals that mark the evaluation of one's actions and person according to the criteria of the day. Even one's feelings come up for moral evaluation. And in addition that social world is full of stories, exemplary anecdotes which reveal the forms of proper and improper social encounters. Each of these features that mark particular social worlds has a part to play in shaping inchoate thought and feeling into ordered minds. In this paper I shall concentrate on a certain class of grammatical models which are not to be found in all languages.

The concept of 'self-consciousness' shares the ambiguity of the word 'self' itself. In the sense in which self-consciousness is a painful awareness that one is being scrutinized by others 'self' seems to mean roughly 'person'. It is a public being who comes under public gaze. But when one speaks of self-consciousness as that which differentiates Men from other primates, who no doubt are conscious of much that goes on around them, we have another sense of 'self' in mind. To be self-conscious in this latter sense is to know that one's mental states are one's own. And to know that one must have a concept of that of which those states are attributes. But the criticisms of the philosophers have put paid to the idea that this kind of self-consciousness is consciousness of self. That which has mental states cannot itself appear as a mental state. The view I wish to advocate in this paper is that to be self-conscious in this sense is to have the ability to play a certain language-game, that of self-ascription, a language-game partly defined by the use of certain grammatical forms.

I have already touched on Wittgenstein's observation that first person mentalistic statements behave very differently from third person statements predicating the very same mental state term. To take his famous example to say 'I am in pain' is not to make an empirical statement about myself for which there might be criteria which might or might not be satisfied, but is part of the *expression* of pain. It is an avowal not an assertion. If I say it when I am not in pain this is a misleading avowal rather than a false description of my inner state. It is more like uttering an insincere promise than it is like telling a lie.

But when I say that someone else is in pain or joyful or thoughtful, and so on, I am ascribing a state, condition, ability, etc. to them on the basis of the satisfaction of certain criteria. Wittgenstein makes use of a distinction between 'criteria' and 'symptoms' to differentiate those conditions which when satisfied necessitate the ascription of the state from those which merely give good reason for the ascription. In some psychological contexts this is an important distinction but it need not detain us in this discussion.

What does this tell us about the point of using the first person? The speech-act 'I can see a tarantula in the hand basin' is not an ascription of a state of seeing to some mysterious inner being, myself. It is an avowal of information, knowledge or belief, indexed to me. It shows how the world looks from where I am standing and who is taking responsibility for introducing this bit of information into the conversation. The statement asserts that the speaker is aware of a tarantula, not that a state of awareness is or pertains to him. A psychologist might be interested in the unattended information processing that preceded the avowal.

If self-consciousness is the capacity to ascribe states to myself as mine then we would expect there to be more complex grammatical forms available by means of which more complex thought forms could be managed. An example of such a form might be 'I think that I can see a tarantula'. Typically such complex sentence forms take epistemic verbs in the first clause, such as 'think', 'know', 'believe' and so on. They serve to express an assessment of the epistemic quality of the avowal made by the use of the embedded clause, in which typically verbs and verbs of action appear. 'I know that I hid the film in the hollow doorknob'. My thesis is that the embedding sentence is also used to make avowals to express my attitude to the embedded avowal. The pronoun 'I' that appears in the first place is an indexical expression for persons, and refers only to the public speaker, that is indexes the epistemic attitude expressed to the embedded avowal as his or hers. But now the speaker is commenting on a speech-act, namely that expressed by the embedded clause. The second part of my thesis is that that speech-act stands to the speaker who produces the whole utterance in just the same relation as he or she would stand to the speech-act of another person. It is as if it were in the third person. But under that condition it has the force of an ascription and so can be considered for its truth value. But to whom or what is the relevant mental state, process, etc. ascribed? It is at this point, I believe, that the concept of 'inner self' gets a purchase. We need it to make this kind of discourse intelligible, since it is this 'self' that records, so to speak, the characteristic unity of each person's mind. It cannot be just another way of referring to the public person, he or she to whom speech-acts are indexed. Conversations could be conducted and avowals exchanged without there being any more to the background cognition than physiological processing. As a matter of fact we experience our thoughts, feelings, recollections, intentions and so on as belonging in one ordered system. To ascribe a mental state, etc. to myself is to locate it in that unified and ordered system. But our grammar demands that there be a subject for any ascription, and the self is that subject.

What sort of concept is the self? By putting the question this way we can avoid, perhaps indefinitely, the need to answer the question 'What sort of thing is the self?'. In the natural sciences there is a class of concepts which seem to perform very much the same role as the 'self' performs in commonsense psychology. These are the theoretical concepts which obey the general grammatical rules of empirical concepts in that they behave like referring expressions, but whose referents are for some reason problematic and are treated as fictional or hypothetical. Quarks, naked singularities, repressed wishes, macro-economic processes, etc. all have this existentially problematic status. Such terms unify scientific discourse by serving as the grammatical subject of ascriptions, allowing us to express the clustering of properties and dispositions into systems. The content of these terms is often created by analogies with the content of terms which do have empirical referents. The logical grammar of the term 'self' is more like that of 'gluon' than it is like that of 'elephant'. To fully understand such terms we must find the source of the analogy by which they get their meaning. I propose that the local concept of 'person' is the source of meaning, through analogy, for local, that is culturally relative concepts of 'self'.

Considered from this point of view, to be a self is not to be a certain kind of being but to be in possession of a certain kind of theory. One uses the theory to organize one's knowledge and experience of one's own states in the way that one uses the physical theories of which one is in command to organise one's knowledge and experience of the physical world. I have suggested that just as the concepts of a physical theory are drawn from analogies with source concepts, so are the concepts of self drawn from analogies with person concepts. The explanation of why selves are unified follows as a trivial consequence of this proposal. Since persons are embodied in such a way that each person has one and only one spatially and temporally continuous body all analogues of the concept of a person, however that concept may differ in different cultures, must preserve the unity. Is 'the self' then just a grammatical feature of how people converse? On my view the self is not the unified structure of mind, but a concept. Nevertheless, the intimate relationship between the possession and use of that concept and the ordered mentation we call minds allows us to explore self-hood through studies of the kinds of structure which that mentation exhibits. We should not mistake this for an empirical study of self. The empirical study of self is made in the public-collective world of conversational conventions and moral orders. The Inuttut language accomplishes person reference with only two devices. The suffix 'ik' directs the hearer's attention to the speaker and functions very much like our pure indexical 'I' in Wittgensteinian avowals. The suffix 'tok' directs the hearer's attention away from the speaker to someone else. But Inuttut does not have a form by which the iterated first person statements through which the speakers of English and many other languages can comment upon the status of their own avowals. On the basis of this feature of Inuttut grammatical models and the extensive ethnographic evidence that seems to show that no other grammatical model for the appraisal of avowals exists, I claim that the Eskimo do not work with our conceptual differentiation between public person and inner self (Harré (1983)). But this is not to lack a component which we possess. They differ from us in the way the holders of geocentric theories of the universe differ from those who believe in heliocentric theories. The rival astronomical theories facilitated different ways of organizing human experience of the visible heavens.

Case 2. Exemplary episodes

But the public-collective conversation also includes exemplary conversational episodes from which models of self-address can be taken. A.J.P. Kenny (1979) has proposed a theory of that which was traditionally called 'the will' in terms of a private, appropriated way of command modelled on public-collective episodes of the exercise of authority. The human will is just the capacity to use an authoritarian form of address directed to oneself. Personal agency is an individual power whose scope and degree depends on the public-collective practices that are available as models for forms of self-command, self-reprimand, and so on.

Case 3. Conventions of discourse

The case of rationality can be used to illustrate the social constructivist principle that conventions of discourse can engender individual mental habits. In this case social constructionism appears as a causal theory by which the existence and properties of an individual mental characteristic is explained by reference to a public-collective practice. This aspect of the theory can be illustrated by two recent pieces of research.

Social historians shown how rapidly accepted beliefs about the rationality of women have changed. Evidence of these changes can be found in the changing psychological presumptions in the laws dealing with various rights of management of property. On the assumption that competent authors do catch the psychological stereotypes of their times we can make cautious use of studies of the portrayal of the psychological capacities of women in plays and novels. In the mid-nineteenth century women are portrayed as enthusiastic, swayed by impulse, and generally irrational in their approach to problems. (See for instance the contrasting proposals offered by Mr. Low and the Duchess of Omnium for bringing about the acquittal of Phineas Finn in his trial for murder.) (cf. A. Trollope, 1874) Can it be that in one short century the cognitive capacity of half the human race has changed so drastically that we can now reasonably demand the same standards of cognitive skill from both men and women? No account of rationality in terms of information processing and brain physiology can possibly be plausible.

A clue to the resolution to this puzzle can be found in some recent studies of the rationality of scientists. The Cartesian picture of mentation requires an inner cognitive 'engine' which processes information according to some logical algorithms, producing such exemplars of rational discourse as printed scientific papers. Microsociologists of science have shown that this picture is quite misleading. Knorr-Cetina's (1981) study of the writing up of scientific papers shows how logical properties are inserted into a discourse in the course of a complex series of social interactions by which community conventions are enforced. In scientific discourse assertions must be backed up by reasons. The scientific community requires that reasons be logically related to conclusions. The question of whether these conventions of discourse can be justified by arguments drawn from philosophy of science is another matter. The production of rational papers by scientists does not depend on, nor does it prove, the existence of rationality as a property of the mentation of individual scientists. One supposes that certain sorts of mental habits will be fostered in a community with so strong a convention as to the form of its public discourse.

Can one identify a conversational convention that would serve as the basis of a social explanation of the alleged irrationality of women? In the sense germane to this paper a speech-act is accountable if the speaker or someone on his or her behalf, could be required to give reasons for what has been said – either as to the very saying of it, or more particularly as to its content. (The other sense of 'accountable' refers to the question of whether what has been said falls or does not fall within the expected repertoire of sayings of a person of that sort.) It is clearly a social convention whether the speech-acts of a conversation are accountable in the first sense, that is whether people are routinely expected to give or to be able to give reasons for what they have said. Women's rationality, according to the social constructivist point of view, is to be understood in terms of changing social conventions of conversation, in particular whether or not the speech-acts of women are accountable. The historical evidence suggests that the conventions of accountability have changed and that in mid-nineteenth century women's speech was not, in general, accountable. Women were not called upon nor expected routinely to provide reasons for their conversational contributions. In these social circumstances the habit of preparing for the request for an account could hardly have been expected to take root. Women's 'intuition' is the mythic cognitive faculty that Cartesianism calls into being 'behind' non-accountable speech, and is no more really an inbuilt property of feminine psychology than would a corresponding 'rationality' be for our contemporary women colleagues. If women have reacted to the changing social conventions governing their speech by acquiring the habit of preparing for demands for accounts, as I suppose those women who take part as of right in accountable conversations no doubt have, so they have acquired 'rationality'.

Once again a feature of processes to be found in my Quadrant III, the realm of the private and individual, has been shown to derive from the grammatical models and typical episode structures of the conversation which fills the public-collective realm, Quadrant I. But why have the conventions of women's speech changed? The socio-economic upheavals of the First World War must have had some importance in altering economic and status relationships. As I have hinted further speculations along these lines ought to be controlled by reference to the sociologies of Marx and Veblen.

K.V. Wilkes has pointed out to me by way of criticism that this treatment of rationality as a discourse-convention engendering a habit of thought, ties rationality rather tightly to language. But what about the rationality of practices? Even in the heyday of women's intuition, wouldn't there have been many expectations of women's rationality of practices which would require an explanation in terms of features of individual cognitive processing? While I would agree that the cognitive processes behind practices including the use of manual skills is a matter of great psychological and philosophical interest, it sows the seeds of confusion to lump where we should split. Rationality has to do paradigmatically, I think, with the giving, the having and the asking for reasons. In this sense, rationality is conversation-bound and clearly related to conventions governing speech and writing. Practical skills routinely involve other assessments and criteria of excellence in performance; for instance they should be efficient, neat, skillful, effective, useful, and they should not be clumsy, time-wasting, and so on. Speaking of the rationality of cooking, child-rearing, and other practices often assigned to women by traditional roles, can only make sense if we have in mind a discourse about these practices in which questions like 'Why do you prick an egg before boiling it?' call for answers that are related in some principled way to the question. 'That's what mother did', or 'It's always done that way' are not answers that can satisfy a rationality-convention.

Case 4. Culturally distinctive language-games

Another way that the social (public-collective) enters into the personal (private-individual) is through the essential involvement of a social component in a concept which is used by an individual in relation to (in interpreting) his or her experience. This case is quite different from (1) in that the social context enters into an individual's assessment of certain features of that context. Thus, as I hope to show, interpreting a feeling as an emotion, an individual must take account both of the phenomenal quality of the feeling and of the context in which the feeling occurs – and in many cases, the context is social. The cognitive processing can be individual (though sometimes it is collective) but draws on both intra- and inter-personal information.

In the case of many emotions an individual's assessment of certain social relations in the context determines, in part, how a feeling is interpreted as an emotion. This can be illustrated with a development and elaboration of Sabini's and Silver's theory of envy and jealousy (J. Sabini and M. Silver, 1982).

B is envious of A if the following conditions obtain:
(1) A has x; (2) A has a right to x; (3) B wants x; (4) B knows or believes (1), (2) and (3).

B is jealous of A if the following conditions obtain:
(1) A has x; (2) B has at least an equal right to x as A; (3) B wants x; (4) B knows or believes (1), (2) and (3).

Differentially under these conditions B will interpret the nasty feeling when he thinks of (sees, etc.) A with x as envy or jealousy – or more probably others will make the judgement for him.

These conditions are not to be read as contingent causal accompaniments of the emotions, but analytical components of the concepts of envy and jealousy. They could be interpreted as rules for the use of the words 'envious' and 'jealous' in a public discourse or in a private discourse derived from it. As such they cannot be disputed by citing 'data' from psychological experiments. They can be upset only by counter-arguments drawn from philosophical (linguistic) analysis.

These conditions reveal themselves in characteristic strategies (which we might use as Wittgensteinian 'symptoms' for assessing the emotional state of someone with whose history and actions we are acquainted). An envious person might challenge the right of the holder to the desired good, or perhaps might denigrate the good itself. The jealous person might try to deprive A of x, in extreme cases by destroying x or killing A.

This is a tentative analysis and no doubt can be further refined by exploring marginal cases, sibling language-games, the extent to which the conditions are severally necessary and/or jointly sufficient for the attribution of the emotional state. The key point for my argument is that in interpreting his feeling as a green emotion, B is taking into account the social relations that obtain between himself and A with respect to the local moral order, that is with respect to the rights each has to some good, status, or advantage. Emotions, or at least some of them, not only involve a characteristic feeling (James-Lange); a characteristic display (Darwin-Leventhal); a hypothesis as to the cause of the feeling (Schachter-Singer), but also a moral assessment dependent on social context. An emotion of this sort is not socially caused, but is socially constructed. It is B's beliefs about rights rather than a public assessment of these rights that is the efficacious element.

A very similar point has been made about anger, reviving Aristotle's classic analysis. To interpret an aggressive and irritated response to something that happens (is done by another, etc.) as anger, the sufferer *must* take the interference to which he or she has been subject as some sort of transgression. This is not a contingent feature of certain happenings to be studied by accumulating experimental data, but a deep grammatical rule governing the language-games of 'anger'. And there are different rules for 'cross', 'irriatated', 'mad', 'furious', 'offended', 'righteously indignant', 'veangeful', and so on across the spectrum of the red emotions.

Heelas has suggested that there may be tribes without emotions. By that he means that a people may have quite other ways than ours of managing, interpreting and classifying the feelings (passions in the literal sense) which are caused by social and physical situations (bereavements, physical dangers, etc.). If those ways lack the causal and moral components which are so prominent in our psychology, then it would be grossly misleading to say that their psychology includes emotions. An important consequence would be that there would be no place in the practices of that culture for the moral assessment of those feelings. For us it is right to feel pity and hope, wrong to feel anger and despair.

Summary

Social constructionism is not a doctrine about how experience in general is possible. It does purport to provide the conceptual basis for explanations of the way experience is organized into individual minds with classifiable and communicable attributes. It opens up the possibility that as societies and their language-games differ, so do the minds whose structures are modelled on features of their sociolinguistic environments. Psychology becomes a science like biology, a science that recognises a diversity of organisms understood as 'solutions' to the 'problems' of the maintenance of life in different environments. In this way it would be very unlike physics, which seeks for underlying universals in the diversity of particular phenomena.

REFERENCES

Dennett, D. *Brainstorms*, Hassocks: Harvester Press, 1979.

Harré, R. *Personal Being*, Oxford: Blackwell, 1983.

Kenny, A.J.P. *Aristotle's Theory of the Will*, New Haven: Yale University Press, 1979.

Knorr-Cetina, K. *The Manufacture of Knowledge*, Oxford: Pergamon Press, 1981.

Kreckel, M. *Communicative Acts and Shared Knowledge in Natural Discourse*, London: Academic Press, 1981.

Mead, G.H. *Mind, Self and Society*, Chicago: Chicago University Press, 1934.

Pearce, W.B. and Cronen, V.E. *Communication, Action and Meaning*, New York: Praeger, 1980.

Rubinstein, D. *Marx and Wittgenstein*, London: Routledge and Kegan Paul, 1981.

Sabini, J. and Silver, M. *Moralities of Everyday Life*, New York: Oxford University Press, 1982.

Schutz, A. and Luckmann, T. *Structures of the Life World*, Evanston: Northwestern University Press, 1973.

Searle, J. *Intentionality*, Cambridge: Cambridge University Press, 1983.

Trollope, A. *Phineas Redux*, London: 1874.

Vygotsky, L. *Thought and Language*, Cambridge, Mass: MIT Press, 1962.

Wittgenstein, L. *Remarks on the Philosophy of Psychology*, Oxford: Blackwell, 1980, vols. I and II.

Part Three

INTENTIONALITY AND RATIONALITY

DAGFINN FØLLESDAL

In this paper, I shall discuss the notions of intentionality and rationality and their bearing on the human studies. I will start with brief analyses of the two notions and the relation between them. I will then concentrate on rationality and devote a substantial part of the paper to the role rationality plays in human studies. Finally, I will discuss the exact status that rationality assumptions should be given in the methodology of the human studies.

1. Intentionality

By "intentionality" is sometimes meant the practical notion of intending to do something. However, beginning with the translations of Aristotle's works from Arabic into Latin in the 12th century, the word "intendere" and related words like "intentio," etc. have taken on another sense, originating in Aristotle's idea that when we perceive an object, our mind takes on the form of the object. This idea was developed further by some of the medieval philosophers and taken up again in the 19th century by Brentano, who transformed it into his thesis of the directedness of the mental. Brentano's view in turn was modified by Husserl and became a key feature of Husserl's phenomenology.

In this paper I shall use the word "intentionality" for this latter, Brentano-Husserl notion of the directedness of the mental. I regard this as the more basic notion in that it is usually presupposed when one explains and discusses intending in the practical sense and rationality.

Let us now make a little more precise what is meant by the directedness of the mental. While for Brentano directedness simply meant that for each mental phenomenon, e.g. for each case of perception, there is some object *towards which* it is directed, *of* or *about* which it is, Husserl had a more discerning view. He acknowledged that many mental phenomena, e.g. hallucinations, do not have any object. Rather than attempting to account for intentionality by appeal to an object that the mental phenomenon is directed towards, Husserl fucused on what the directedness consists of: what are the features of the

Margolis, J., Krausz, M. and Burian, R.M., (eds.), Rationality, Relativism and the Human Sciences.
 ISBN 90-247-3271-9.

mental thanks to which it is always *as if* it has an object? Husserl called the collection of these features the *noema*. He regarded the noema as a generalized notion of meaning, thereby tying together inten*t*ion with a *t* and inten*s*ion with an *s*.

I shall not go into Husserl's particular analysis here. I will focus only on one special feature on the analysis which is of importance in connection with rationality: according to Husserl, the directedness that characterizes consciousness, consists of our extrapolating beyond what presently meets the eye. To experience an object is to experience something with a back, that is presently out of view, and also with a past and a future. Even the features of the object that presently meet the eye are features only because they have a place within an elaborate totality where perception, imagination and action are woven together.

Whenever we are conscious of an object, we regard it as having a place within such an extrapolated totality. This totality we call the world. We have to distinguish on the one side our *representation* of the object with the world around it (this representation is what Husserl called the noema), and on the other side the *object* and the surrounding world. The object and the world are largely unknown. Some of their features are precisely and firmly anticipated in the noema, other of their features are only anticipated in rough outline or not anticipated at all. Our noemata are hence open-ended in at least three ways: The world is regarded as containing lots of objects that we do not know about and perhaps have no inkling about, most of the properties of these objects and the relations between them are unknown, and thirdly, even most of the properties of the object that we are presently focusing our attention on go beyond our anticipation.

In addition to this open-endedness there is our fallibility. Even where we have pretty firm beliefs about the existence of objects, about their properties and interrelations, we may be wrong. This is a reason why Husserl wanted an analysis of intentionality that does not presuppose that there is always an object where there is intentionality.

A major problem in Husserl's phenomenology is how we can find out anything about the noema and what the identity criteria of these entities are, to use Quine's terminology. Husserl is aware of the problem and discusses it at length. He developed what he called "the phenomenological reduction" as a method for studying the noemata. I do not find this method satisfactory. However, I find Husserl more incisive than many contemporary philosophers who develop theories of mental representation but do not even seem to be aware of the problems.

I shall discuss this problem briefly later. The problem is important, since intentionality, whether it be analyzed by appeal to a noema or in other ways, seems to be presupposed by rationality, to which we now turn.

2. Rationality

Rationality seems to be an even more multifarious notion than intentionality. The literature abounds with different and often seemingly unrelated notions of rationality, from various kinds of "minimal rationality" to stronger notions. To exemplify one of the stronger notions it is enough to remind you of Rawls: "The rational individual does not suffer from envy" (*A Theory of Justice*, p. 143).

My colleague in Oslo, Jon Elster, has written a survey article on rationality for a forthcoming volume published by the Institute International de Philosophie in which he distinguishes more than 20 senses of rationality. These are not all in conflict – they often relate to different areas: rational behavior (rationality as efficiency), rational beliefs (rationality as consistency), etc.

In this paper, I will concentrate on rational behavior, but I will also make some remarks on rational beliefs and rational values, and I will emphasize a distinction that Elster does not make, between normative and descriptive theory of rationality.

Let me first, however, mention briefly four main kinds of rationality and indicate how they are connected with intentionality. They are (1) rationality as logical consistency, (2) rationality as well-foundedness of beliefs, (3) rationality as well-foundedness of values, and (4) rationality of action.

(1) Rationality as logical consistency is the requirement that a person's beliefs shall be logically consistent with one another.

This may mean several different things, depending on two factors that may vary independently: First, how much are we to incude in a person's beliefs? Three options that immediately come to mind are (i) the beliefs we actively entertain in a given moment, i.e. those that we are presently "thinking about," (ii) those beliefs that play a role in determining the actions, if any, that we are presently carrying out (we are going to return to this in connection with rationality of action later), (iii) those beliefs that can be elicited by questioning a person, as for example in the questioning of the slave boy in Plato's *Menon*.

Note that while group (ii) presumably includes group (i), group (ii) and (iii) may be incomparable. Group (iii) obviously would include lots of beliefs that are not included in group (ii), many mathematical truths that are not yet even conjectured would, for example, belong in group (iii). Group (ii), however, in its turn, appears to include many beliefs that are not included in group (iii). If Freud was right, each of us has many such beliefs which influence our actions, but which cannot be elicited by normal questioning.

The second factor that is of importance for what is meant by our beliefs being logically consistent is whether we just mean that none of our beliefs is the negation of another, or whether we mean that the set of our beliefs does not logically imply a contradiction. The latter condition is obviously stronger, and perhaps none of us satisfies this rationality condition.

Husserl's noema corresponds fairly well to the set of our beliefs in group (ii) above, those that play a role in determining our actions, but not all logical consequences of these beliefs. The noema is, as you remember, open-ended.

According to Husserl, the noema, as far as it extends, does not contain contradictory features, although we may upon reflection come to see that a noema we had at a given time, e.g. that of a largest prime number, does imply a contradiction. The noema hence is logically consistent in the weak sense of logical consistency that I just mentioned. Rationality in this sense, as weak consistency of the set of beliefs that determines our actions, is therefore part of the general Husserlian theory of intentionality: our experience of the world, including our anticipation of it, contains no contradictory features.

The analysis of intentionality in terms of noemata is therefore not equivalent to an analysis in terms of "possible worlds." The fact that a noema does not contain any contradictory features does not guarantee that there is a possible world that is compatible with it; noemata do not invariably permit consistent extensions to a complete "world."

(2) The second kind of rationality I mentioned was rationality as well-foundedness of beliefs. This is a much stronger notion than the previous one. That our beliefs do not contradict one another is necessary, but by no means sufficient for their being well-founded. Well-foundedness requires something more, namely that our beliefs be well supported by the available evidence. Our beliefs may go well beyond the available evidence, as they do in the case of the more theoretical parts of scientific theories. But there should be no other competing theories that would be better supported by the available evidence. The specification of well-foundedness would recapitulate epistemology and scientific methodology; we should only note that such a specification must also make precise the phrase "available evidence." Rationality as well-foundedness concerns not only what beliefs we should hold *given* a certain amount of evidence, but also to what extent it is rational to search actively for additional evidence before we allow our beliefs to become fixed.

Rationality as well-foundedness of belief is clearly a normative notion, not a descriptive one; most of us are not very rational in this sense most of the time. I shall come back to this normative-descriptive issue later. I shall also discuss further the relation of this kind of rationality to intentionality; since we are not always rational in this sense, but presumably always manifest intentionality, this kind of rationality is not required for intentionality.

(3) Rationality as well-foundedness of values. A doctrine that has been infuential since Max Weber, is that while one may choose means towards an end in a more or less rational way, there is no notion of rationality that applies to the evaluation of ends, or values. This view does not seem to me to be correct. I think that our normative considerations can also be more or less

rational and that normative judgments, like factual judgments, can be founded with more or less rationality.

The most promising approach to this issue is in my view the method of a reflective equilibrium, that has been worked out by Goodman, Rawls, and others. I shall not go into this view here, since I assume that you all know it. What is important from the perspective of our discussion is that this view on justification of normative statements gives us a way to judge the rationality of normative statements. Rationality in ethics become similar to rationality in science: the rationality, or well-foundedness, of our judgments depends upon the degree to which we have achieved a reflective equilibrium in which our general principles fit in with one another and with our adjusted experiences.

As in the case of rationality as well-foundedness of beliefs, rationality as well-foundedness of values is a normative notion, not a descriptive one, and it is not required for intentionality. I shall come back to this issue later.

(4) The fourth and last of the kinds of rationality that I wanted to discuss was rationality of action. Since I have discussed rationality of action more thoroughly in a couple of earlier papers, I shall be very brief here. I will only say that I find decision theory, as it has been developed by economists, mathematicians, and philosophers, the best framework currently available for discussing the rationality of actions.

According to this theory, an action is the end product of a two-step process. First we consider which actions are open to us in our given situation. This is largely a matter of our beliefs concerning what is possible for us and what not. In praxis, however, we do not survey all possible alternatives, our search is limited partly by our phantasy and the time at our disposal, and partly also by our conception of how much of a bearing the various alternatives have on what we desire or fear. We tend to focus on alternatives that seem to be particularly important for us to realize or avoid, i.e. alternatives with particularly great positive or negative expected utility, as the economists say.

The second step in the process now consists in weighing the envisaged alternatives against one another, taking into account both our beliefs concerning the probabilities of their various consequences and the values that we assign to each of these consequences. Multiplying the probabilities with the values and adding up, we then arrive at the expected utility of each alternative, whereupon we choose the alternative with the highest expected utility. This multiplication and addition is subject to severe difficulties which have been extensively discussed in the literature of decision theory. The way in which one pools preferences and values into a resultant expected utility may turn out to have little to do with the arithmetical operations of multiplication and addition.

One comment may be in place here: Some versions of decision theory treat the set of alternatives and consequences in a platonistic way, as the set of all physically possible alternatives open to the agent in the given situation and all physically possible consequences of them. Clearly, this is not the appropriate conception if we want to clarify rationality. Rationality always has to do with what the agent ought to choose, given his or her limited, perspectival view on the situation, with a limited amount of information, limited phantasy and time for considering different alternatives and thinking through their various consequences. Rationality of action is normally a question of how to make the best of one's resources, one's information-seeking capabilities, and one's ability to create good alternatives, and not a question of choosing from within a vast set of alternatives that lie there ready for one's inspection.

Even when understood in this weak, practical sense, rationality of action may be a normative rather than a descriptive notion. If so, it is clearly not required for intentionality. We shall return to this question later. Note, however, how rationality of action avails itself of representations or noemata: The agent imagines, or represents to himself, several different actions and ensuing states of the world and tries to choose between them. I mentioned earlier how the notion of representations or noemata raises serious epistemological problems. However, without such notions, it seems very difficult to get an adequate theory of action.

3. The role of intentionality in human studies

Given that intentionality characterizes all consciousness, intentionality can be expected to pervade the study of consciousness and of its products, i.e. the human studies. The main implications of this for the human studies are the following:

(1) SUBJECT MATTER

In the human studies we are not primarily interested in what the world is like, as we are in the study of Nature, including anatomy and somatic medicine. In the human studies we are interested in what a person's noema, or representation of the world, is like. We are interested in his or her perspective on the world, way of experiencing the world.

(2) METHOD

The human studies require a method for studying representations or noemata. Philosophers who speak of representations are often remarkably quiet when it comes to how we can know something about them. Husserl developed the phenomenological method for this purpose, a method whereby one studies the noemata directly, through a special kind of reflection. This method is often propounded as the appropriate method for the human studies.

In spite of Husserl's acute observations about intentionality and our ways of getting at it, I am, however, not satisfied with his method. Instead I will argue that a person's representation of the world, his or her beliefs and values, must be studied in a more roundabout way. My view is that the hypothetico-deductive method that is used in the natural sciences must be used in the human studies as well. However, one of the major differences between the natural sciences and the human studies is that in the human studies one makes assumptions about rationality. These assumptions are very basic hypotheses that may seem necessary in order to set the human studies going. We will return to this question later.

Nevertheless, as I see it, intentionality and rationality are intimately connected in the following way: *Rationality opens the way to the study of intentionality*. This connection is of crucial importance for the human studies.

In the remainder of this paper I shall briefly sketch how rationality assumptions enter into the study of intentionality. Much of what I say is inspired by Quine, and much of it will be familiar to you from the writings of Donald Davidson. However, there are some differences between Davidson and myself, in particular when it comes to the status of the rationality assumptions, which I will discuss at the end of the paper.

4. The role of rationality in human studies

The main aim of the human studies is, as I mentioned briefly in the preceding section, to understand a person's noema, or representation of the world, i.e., his beliefs and values. Two related aims are to understand his actions, which as I indicated above, depend upon these beliefs and values, and to understand the results of these actions, to the extent that these results manifest the agent's beliefs and values. Written texts, speech and works of art are typical such manifestations, but also various social institutions, etc., which have been brought about through an interplay between several agents.

Now, the simplest way to go would be to start with the person's beliefs and values and proceed to the actions and results of actions from there. This would be the natural way to go if we had direct access to a person's beliefs and values. However, in my opinion we do not have such access to other people's beliefs and values. And even if we should have such direct access to some of our own beliefs and values, there would still remain the problem of communicating our self-insight to others. Husserl does not seem to be aware of the seriousness of this problem. Communication, I maintain, is not based on a direct reading off of the other's beliefs and values, but on a starting out from evidence that reaches us through our senses and constructing hypotheses about the other's beliefs and values.

This is where rationality assumptions come in. In trying to understand the other person's actions I attribute beliefs and values to him on the assumption that he is rational. In trying to understand what the other person says or writes, I cannot take it for granted that he means by the words what I mean. Therefore, I cannot attribute to him the beliefs I would have if I assented to the same sentences. In the learning of the semantic aspects of language, belief and meaning are intertwined in such a way that there are not two elements there to be separated. This is an ontological point, not just an epistemological one. That is, the point is not that there are two elements there, meaning and belief, which we are unable to separate in a unique way, given the limited evidence we have available. The point is rather that meaning is nothing but a complement to belief. If we attribute one set of beliefs to a person, we come to interpret his sentences one way; if we attribute another set of beliefs to him, we come to interpret his sentences another way. This is one way of putting Quine's doctrine of indeterminacy of translation, the way I interpret him.

Now, if beliefs in turn were nothing but a complement to meaning, there would be nothing to be right or wrong about in the whole realm of belief and meaning. However, this is not so. I regard beliefs as dispositions which are instilled in us through perception and inference and which manifest themselves in action. By starting from perception and action we can therefore work ourselves back to a partial knowledge of a person's beliefs and values. This is where intentionality and rationality come in.

To take perception first, in a paper in 1974[1] I argued that perception cannot be dealt with by way exclusively of the notion of stimulus the way Quine does in *Word and Object*. Perception manifests intentionality, it contains a representative element, a noema, a set of beliefs concerning what goes beyond the mere stimulation. As Husserl pointed out, although everything may be the same in my physical surroundings and on my sensory surfaces, what I experience, what I see, may vary greatly. My verbal utterances and my actions depend on what I perceive and not on what I receive. This creates difficulties for the version of the behaviorism that Quine expounded in *Word and Object*. The difference between perception and reception is discussed by Quine in *The Roots of Reference*, where he tries to deal with this distinction without introducing intentionality.

Whatever view we may have on perception, the following is the basic methodological point that bears on the human studies: One should attribute to a person the beliefs that one considers it most likely that that person will have, given one's epistemology and given one's knowledge of the person's experiences, education, mental powers and actions.

Similar considerations come in when we construct hypotheses concerning what normative views and values a person has. We use our theory of rationality in ethics, e.g. a theory of reflective equilibrium, and again take into account what we know about the person's experiences, education, upbringing, reflective inclinations, and actions.

Note how actions come in in both cases, both in connection with beliefs and in connection with values. Observation of action is a major source of evidence for our hypotheses concerning a person's beliefs and values, since both beliefs and values play a role in explaining a person's actions. Again we assume that the person is rational and then try to find out what beliefs and values we may attribute to him that are compatible with his being rational.

Before I go on to discuss this rationality assumption further, let me make a short remark about the observability of actions. Some behaviorists hold that I do not see actions, I see only bodily movements, and then *infer* that they are actions. Following Husserl, I regard this as wrong. I see actions, but I should be aware that in seeing an action, as in seeing anything else, there is an hypothetical element. If my theories of the world or my understanding of the person I have in front of me were to change, I might come to see a quite different action where I now see this one, or I might even come to see only a bodily movement. When an economist reports that a person acts in such and such a way, which supports one of the economist's theories, he might later come to see that what the person did was something quite different, that from a mere physical point of view looked very much the same.

It has long been popular among economists studying so-called "revealed preference" to hold that the only way of determining a person's beliefs and values is to examine his actual choices, there is no non-choice source of information concerning a person's beliefs and values. This gives us a very small circle. We explain a person's choices by appeal to his preferences, and we attribute beliefs and values to him on the basic of his choices.

Now we clearly have other sources of information. We can ask him about his beliefs and values. However, there are difficulties here. First, we can not always trust what a person says. He may give us a false story of what his reasons are for acting in the way he does. He may also lie to himself, he may rationalize. His behavior may be due to reasons that he does not know. It may be due to factors that might call for a Freudian or a physiological explanation. Thus, to use an example given by Patrick Suppes, a young boy who has just entered puberty and has an attractive female teacher may very frequently come up to the teacher after class to ask questions concerning his school work. When asked why he does this, he may answer that he has these questions concerning the school work and that he wants to learn. This may be his sincere answer, but we may want to give a different explanation of his actions.

Note, by the way, that just as our theory of explanation of actions must have room for such deviant phenomena, so on the other hand the classification of something as deviant, as rationalization, as repression, sublimation, etc., is possible only on the basis of such a theory of how actions should be explained.

It is also highly important to remember also that the information we get from a person by asking him questions, listening to what he says, etc. is ultimately based upon his behavior, notably his linguistic behavior. Asking him questions concerning his reasons for acting as he did, broadens our base from comprising just choice-behavior to comprising behavior in general. We are therefore still caught in a circle, we explain a person's behavior in terms of his values and beliefs, which in turn are attributed to him only on the basis of his behavior.

As in the case of all circular explanation, we have a lot of leeway here. It is, for example, well known that when we explain a person's actions, we attribute different values to him depending on what beliefs we take him to have, and conversely different assumptions concerning his values will yield different conclusions concerning his beliefs. Davidson has pointed out that this interdependence between values and beliefs is similar to the interdependence that we noted earlier, in connection with Quine's thesis of indeterminacy of translation, between what we take a person's utterances to mean and what beliefs we attribute to him.[2]

Fortunately, there are various restrictions here, that help to cut down on all this under-determinateness. One such restriction, which is quite obvious, is due to the fact that the two interdependent pairs that we just considered, beliefs and values in the case of action and beliefs and meaning in the case of utterances, have one member in common, viz. belief. Clearly, the beliefs that we attribute to a person when we seek to explain his actions have to fit into the same belief system as the beliefs that we attribute to him where we try to interpret what he says.

Other restrictions, that I have discussed elsewhere[3] have to do with perception and with ostension. Further restrictions emerge when we give a detailed analysis of speech acts as a species of actions and incorporate into our interpretation of the other person the assumptions we make about his beliefs and values when we seek to explain his speech acts.

5. Some refinements on the notion of rationality

The decision-theoretic pattern of explanation of action that I have just sketched is often regarded as the paradigm of rationality, the model of rational decision making. One always chooses what is best for one, or more accurately, what one believes to be best for one.

In proposing this pattern as a basis for explaining actions and thereby for understanding them, I admit I know no better theory. Some theory of action based on rationality is, I think, necessary in order to start the enterprise of explaining and understanding actions. I shall return to this in my conclusion. However, although in this sense such a theory is a presupposition of understanding, I do not regard this particular theory as in any way necessary. It is a working hypothesis, which I am willing to replace by some other hypothesis as soon as something better comes along.

In any case, the pattern has to be refined and extended in several ways. We shall now consider five such ways:[4]

(i) *Consistency at a time*

Although the decision-theoretic pattern of explanation is often called the rational choice theory of human behavior, almost any agent could be made to fit into this pattern without seeming very rational. Trying to determine a person's preferences, we may for example find that he prefers a to b, and b to c, but c to a. This seems irrational – and also difficult to understand, as we should expect if assumptions about rationality seem to be a prerequisite for understanding, as I have claimed. It is common therefore, to require for rationality also the following: a rational individual who at time t is placed between a number of alternatives to be realized makes his choice according to complete and transitive preferences at t. If a person satisfies this condition and also the various others that we have been and will be discussing, then we feel that we understand him; if not, we do not yet understand him and we must continue our search for an explanation of his behavior, perhaps by no longer regarding the behavior as action but as something that in fact springs from psychological or neurological causes and calls for physical explanation.

(ii) *Consistency over time*

There are further conditions that have to be imposed upon our interpretation of a person's behavior. The rationality condition I just mentioned concerned only a person's perferences at a given instant of time t. He could satisfy such a condition at every instant of time and yet appear quite irrational and hence difficult to understand. Thus, for example, given a choice between three options a, b, and c he might first prefer a, in the next moment b and then immediately after c. This kind of inconstancy would tend to disturb us. We expect a person's preferences to be consistent not only at a given time, but also over time. This does not mean that we expect a person never to change his preferences, but we would like to understand why he changes them. We would like to have an explanation, by reasons or by causes, for such changes; at other times we may want explanations for lack of changes. This, by the way, is an example that illustrated how an appeal to reasoning and information acquisition, e.g. through perception, enters into the explanation and understanding of action. There are numerous such ties between a person's actions and the other features of a person we want to understand, such as his thoughts, beliefs, feelings, fears, and other mental states.

(iii) *Concern for the future*

Further, we expect a rational person to let his preferences be guided not only by his present desires but also his future ones. Some concern for our own future seems to be part of rationality. This means that it would be irrational to discount completely our future. It would, however, also be irrational not to discount it at all, given human mortality and the uncertainties of life. Hence both very rapid discounting of the future and very slow discounting are phenomena that may defy understanding and call for special explanation.

(iv) *Interaction between agents*

In order to understand human action, we must take into account that many of the beliefs that guide our actions are beliefs about the beliefs and values of other persons and about how they will act as a result of our acts. The standard framework for the study of such interaction between agents is game theory, which could also be called interaction theory, in order to avoid the misleading association many have in connection with the word ''game''. Since so many of the human actions that we seek to understand are just cases of such interaction, it seems to me that no satisfactory study of man can take place without game theory. Game theory is one of the important tools of the human studies. Many cases of group behavior that initially may seem puzzling and irrational, e.g. people throwing litter and Chinese peasants cutting down the forests and thereby bringing about erosion, get simple, rational explanations in terms of game theory, and we thereby come to *understand* them.

Game theory also seems to me to be an excellent framework for analysing speech acts and for determining the various beliefs and values that may go into such acts. However, this is a topic for itself, which I shall not enter here.

6. The status of rationality assumptions

I will end my paper with some reflections on the status of the rationality assumptions we make when we try to understand man, his actions and the results of these actions. Philosophers differ widely in the status they give to the statement that man is rational.

Aristotle maintained that to be rational is definitory of man; man is a rational animal. In our time, Donals Davidson is one of those who most vigorously as argued that in order to understand man and attribute beliefs, desires and actions to him, we have to assume that he is rational:

> . . . the satisfaction of conditions of consistency and rational coherence may be viewed as constitutive of the range of application of such concepts as those of belief, desire, intention and action.[5]

> . . . if we are intelligibly to attribute attitudes and beliefs, or usefully to describe motions as behavior, then we are committed to finding, in the pattern of behavior, belief and desire, a large degree of rationality and consistency.[6]

Similarly, William H. Dray argues:

> Understanding is achieved when the historian can see the reasonableness of a man's doing what this agent did, given the beliefs and purposes he referred to (what the agent believed to be the facts of his situation, including the likely results of taking various courses of action considered open to him, and what he wanted to accomplish: his purposes, goals, or motives).[7]

While Davidson regards rationality as necessary for the very applicability of concepts like belief, desire, intention and action, Dray considers rationality as necessary at least for our knowing what the other's beliefs, purposes, goals or motives are.

Carl G. Hempel, on the other hand, disagrees with Dray and regards the assumption that man is rational as merely an empirical hypothesis, which presumably may be false:

> Now, information to the effect that the agent A was in a situation of kind C, and that in such a situation the rational thing to do is x, affords grounds for believing that it would have been rational for A to do x; but for not believing that A did in fact do x. To justify this latter belief, we clearly need a further explanatory assumption, namely that – at least at the time in question – A was a rational agent and thus was disposed to do whatever was rational under the circumstances.
>
> But when this assumption is added, the answer to the question "Why did A do x?" takes the following form:
>
> A was in a situation of type C
> A was a rational agent
> In a situation of type C any rational agent will do x

Therefore A did x[8]

Hempel hence seems to regard the assumption that the agent is rational as an empirical premise which enters into explanation of action and which must be true if the explanation is to be correct, but which may conceivably be false, in which case our explanation of the person's action is faulty. A similar view is argued for, with more precision and detail, by Wolfgang Stegmüller, in *Probleme und Resultate des Wissenschaftstheorie und Analytischen Philosophie.*[9]

Karl Popper maintains that no rationality assumption is needed:

> the method of applying a situational logic . . . is not based on any psychological assumption concerning the rationality (or otherwise) of "human nature".[10]

Alan Donagan, finally, regards the assumption of rationality as plainly false (and presumably as not required for an explanation of action):

> There is no reason to believe that all historical agents are rational, in any of the several senses of "rational" which Hempel has explored.[11]

> Considering what human history has been, an historian would be in a pretty pass if he were obliged to assume that the only actions he may succeed in understanding were rational. They must, indeed, be intelligible, but that is another thing.[12]

It is my own view that man is not rational in this sense. The assumption that he is, goes too strongly against the evidence we get by observing human behavior, our own as well as others'. Admittedly, we have a lot of freedom in attributing beliefs and values to a person in such a way as to make what he does and says come out rational. As Tversky and Davidson have pointed out,[13] normative decision theory may, for example, be made compatible with many kinds of behavior, since to quote Tversky, "the axioms of utility theory can be regarded as maxims of rational choice only in conjunction with an intended interpretation and the criteria for the selection of an interpretation are not part of utility theory."[14]

However, we are not completely free to make whatever assumptions we want concerning what alternatives and consequences an agent has taken into consideration. Our assumptions have to be reasonable, i.e. they must fit in with what we should expect on the basis of our knowledge about the agent, his beliefs, his imaginativeness, but past experience and performance, the possible influence of panic, pressure, etc. on his decision process, and so on. The same holds for our assumptions concerning his beliefs and values. They must accord with our theory of how beliefs and attitudes are formed and changed under the influence of experience, reflection, etc., together with the information we have concerning his past experience. All of these considerations, too, involve assumptions about rationality, about what beliefs and values a rational person can be expected to have, given his experiences.

In trying to understand a person and his actions, we must weigh against one another all these considerations that we make on the basis of observation of his actions and what goes on at his sensory surfaces, listening to what he says, etc. While a normative theory of rationality provides a framework for all of these considerations, causal, irrational factors come into the picture at a number of places, where our general theory of man makes us expect them to come in, as when a person gets his beliefs and values formed through propaganda, advertizing or group pressure, or when he acts under the influence of hypnosis, drugs, drives with which he is not yet familiar, etc.

All these restrictions and further considerations make it extremely difficult to interpret all that a person does so that he comes out completely rational. In my view, the rationality assumption is not necessary, as Davidson and Dray argue in the passages I quoted, nor is it a true empirical hypothesis. It is simply false.

One way out would now be to regard the rationality assumption as a normative methodological hypothesis in the sense that in attempting to understand man we should always try as hard as possible to make him come out rational. Only as a last resort, when we cannot manage to make him come out rational regardless of what beliefs and values we attribute to him, should we treat his behavior as irrational. There is much to be said for this view. However, I think it is wrong, but less radically wrong than the first view I discussed. Suppes' example concerning the boy in early puberty is a case which shows this second view to be wrong. If we should always do our best to get a person to come out rational, we should accept the reasons this young boy gives for his going up to the teacher. However, in view of our knowledge of psychology we may opt for a different explanation of his action.

What then, is the status of rationality assumptions in the study of man and intentionality?

I will propose a third view: *man has rationality as a norm*. Man is not always rational, nor should we always and at all costs try to regard man as rational. But we should regard man as always being inclined to mend his ways towards more rationality when his lack of rationality is pointed out to him in terms that he can understand. We may, for example, like Aristotle in the Nichomachean ethics, be misled into inferring from "everything is striving towards an end" to "there is an end that everything is striving towards." (The fact that one may easily come to accept such an inference is, by the way, enough to refute both the first two views on rationality that we discussed, that man is invariably rational, and that man ought to be regarded as invariably rational.) However, when it is pointed out to us that this inference is parallel to the inference from "everybody has a mother" to "there is a mother that everybody has," then we are likely not to trust the first inference any longer. There are differences between people in how much detail they need before they give up their irrationalities. However, this is my thesis, that man has rationality as a second-order disposition, a disposition, given appropriate circumstances, to adjust his beliefs and values and actions in such a way as to make them more rational. It is rationality in this sense that should be assumed when we study man and intentionality. We should assume that man has rational beliefs and values and acts rationally unless we can show that his second-order disposition for rationality has been outweighed by other factors, like the complexity of the subject matter or the situation, or other causal factors.

This second-order disposition for rationality, like many of the second-order dispositions we have, may be a product of evolution. Individuals lacking this disposition, and generally the disposition to extrapolate in appropriate ways from past and present experience to the future, may never have reached procreational age.

My thesis, then, is that man is a rational animal in the sense that *man has rationality as a norm*.

NOTES

1 "Meaning and experience." In Samual Gutterplan (ed.), *Mind and Language: Wolfson College Lectures 1974*. Oxford : Clarendon Press, 1975.
2 "Radical interpretation." *Dialectica 27* (1973), 313–327.
3 "Meaning and experience."
4 In the following, I am indebted to several articles and books by Jon Elster, especially his *Ulysses and the Sirens: Studies in Rationality and Irrationality*. Cambridge: Cambridge University Press, 1979.
5 Donald Davidson, "Psychology as Philosophy," p. 237 of the reprint in Donald Davidson, *Essays on Actions and Events*. Oxford: Clarendon Press, 1980.
6 Donald Davidson, *loc. cit.*
7 William H. Dray, "The Historical Explanation of Actions Reconsidered," p. 106 of the reprint in William H. Dray (ed.), *Philosophical Analysis and History*. New York: Harper & Row, 1966.
8 Carl G. Hempel, "Rational Action" *Proceedings and Addresses of the American Philosophical Association 1961–1962*, Vol. XXXV, October, 1962. Yellow Springs, Ohio: The Antioch Press, 1962, p. 12.
9 Wolfgang Stegmüller, *Probleme und Resultate der Wissenschaftstheorie und Analytischen Philosophie*. Berlin: Springer, 1973.
10 Karl Popper, *The Open Society and Its Enemies* (3rd edition). London: Routledge and Kegan Paul, 1957, vol. II, p. 97. Cited in Alan Donagan, "The Popper-Hempel Theory Reconsidered," *History and Theory* 4 (1964), 3–26; pages 147–148 of the reprint of this article in Dray (ed.) *Philosophical Analysis and History*.

11 Alan Donagan, *op. cit.*, p. 155 of the reprint in Dray.
12 Alan Donagan, *loc. cit.*
13 Amos Tversky, "A critique of expected utility theory: descriptive and normative considerations." *Erkenntnis 9* (1975), 163–173. Donald Davidson, "Hempel on explaining action." *Erkenntnis 10* (1976), 239–253.
14 Tversky, *op. cit.*, p. 172.

THE RATIONALITY OF SCIENCE[1]

HUGH LACEY

I

Modern Western culture understands itself as the foremost bearer of rationality, and this self-understanding rests upon the twin pillars of science and technology. From one perspective, that in which *knowledge claims* are recognized as the primary location for rational evaluation, science looms larger. In its light the posits of scientific theory, more than any other form of understanding, gain support from rational evaluative canons. They offer the best account we have of the nature and ways of the physical world, and consequently provide the theoretical underpinning of technological success and advance. From another perspective, that in which rational evaluation pertains in the first instance to *actions* designed to enhance the exercise of our designs upon the world, technology is in the foreground. Then science gains rational precedence as a form of understanding because it provides the theory that furthers technological practice.

Movement back and forth between the two perspectives is easy and common, since they are mutually reinforcing: the hegemony of technological practice is often grounded on its being informed by scientific theory which offers a superior understanding of the world; the pursuit of science and the massive material investment in it are often legitimized in terms of its contribution to technological development. Either way, whether technological possibilities arise as a consequence of science's superior understanding of the physical world, or the superiority of scientific understanding derives from its informing technological practice, the appeal to science carries an unmatched authority. Science emerges as the exemplar of rationality, of rational activity, rational discourse and rational judgement. Commitment to rationality requires a way of life informed by, or at least aspiring to be informed by the claims of the currently best established scientific theories, and a culture which fosters the pursuit of science bears a superior rationality. So runs a common wisdom.

The common wisdom has not been without its critics, and they have multiplied in recent years. Yet its pull upon us remains strong, and we resist

Margolis, J., Krausz, M. and Burian, R.M., (eds.), Rationality, Relativism and the Human Sciences.
 ISBN 90-247-3271-9.

encroachments upon it suspecting them of harboring irrationality, so central is it to the self-understanding of our culture. The critics draw from two main sources, reflections on the history of science and on anthropology, and they often espouse various relativist theses, since their criticisms tend to involve the denial that there are standards of rationality which are not bounded by culture, epoch and society. Nowadays in certain liberal circles relativism is celebrated, being seen as an antidote to the Western bad habit of imposing ethnocentric categories on alien cultures and previous times.[2] Even so, whatever its political merits might be thought to be, relativism tends to be hounded by well-known paradoxes. Thus, a view which is sensitive to the arguments of these critics but not implicated in relativism would have obvious merits.

II

While the common wisdom affirms that science is the bearer of a superior rationality, the philosophical tradition, despite 300 years of trying, has not produced sound, clear-cut arguments to support it. For the most part, the tradition has assumed the first perspective stated at the outset, that the primary location of rationality lies in the account of the physical world posited in scientific theory which, it has supposed, can be established independently of any consequences for technological application.[3] Many have supposed that the relevant arguments to demonstrate the superior rationality of science would rest upon the exhibition of rule-like criteria, derived *a priori*, whose application would settle quasi-algorithmically which of competing scientific theories are superior, and also make clear that modern scientific theories are superior to their predecessors (and to any competing forms of understanding developed by other cultures) in their account of the physical world.

Anthropologists have repeatedly objected that the various rule-like criteria that have been proposed are neither *a priori* nor culturally neutral, but merely reflections of the interests and standards of our culture,[4] with no inherent claim upon the participants in another culture. So, they objected, the affirmation of the superior rationality of science is merely the projection onto human nature and human culture universally of the defining characteristics of our local (Western, for the past 3 or 4 centuries) self-understanding. For the most part, the philosophical tradition remained unmoved by these objections, charging that the anthropologists had made the mistake of inferring from the *de facto* apparent absence of shared standards for assessing judgements about the physical world that there are no criteria of assessment derived *a priori*. After all, part of the lesser rationality of an alien or earlier culture may be its failure to utilize the criteria derived *a priori*.

Exchanges of this type can easily degenerate into posturing, especially if the rule-like criteria and their *a priori* derivations are not actually exhibited. Here there is a problem; the appropriate rules have not been forthcoming. The consensus of the mainstream of the tradition is that the appropriate rules will select among theories (and derivatively of forms of understanding) on the basis of their logical – deductive or inductive – relations with empirical data.[5] As is well known, there has been no shortage of candidates for such rules – e.g., rules derived from inductive logic (either Baconian[6] or Bayesian[7]), from the alleged hypothetico-deductive structure of scientific theories,[8] from the subtle interaction of deductive and inductive moves,[9] or from the methodology of research programmes.[10]

Every one of these candidates has an initial plausibility, and is exemplified in some key scientific episode. However, the results of their applications in practice tend to vary with interpretations of the empirical data, so that they cannot be applied unambiguously unless accompanied by conventions which delineate the relevant class and descriptive categories of the empirical data. But theoretical disagreement in certain crucial cases is inextricably allied with disagreement about the appropriate convention, thereby rendering any proposed rule a secondary matter.[11] In addition, there is the problem that no consistent subset of rules has been deployed plausibly to explain all the cases of theoretical choice that have been accepted as definitive within the canons of scientific practice (even after "rational reconstruction.")

All of this adds up to a crisis in attempts to account for the rationality of science in terms that ground it upon rules of the kind under consideration. We seem to be driven to the conclusion that if scientific rationality is to be explicable in terms of theory choice being rule-governed, then the relevant rules are yet to be discovered, or that if some subset of the above rules is justified *a priori*, then much of what is recognized as scientific progress cannot be explained as generated or vindicated by such rules. On neither alternative *has* the rationality of science been explained. This conclusion has become commonplace in recent philosophical reflections upon the history of science.[12] I concur with it, but will not defend it here. Rather I want to enter into the dialectical situation created by acceptance of it.

III

A number of options appear to be open for the next move. One might be tempted to reject the claim that science bears a higher rationality, and thence the common wisdom stated at the outset. In his more extravagant moments, Feyerabend seems to have given way to this temptation. More commonly, we find that the common wisdom is firmly entrenched, more firmly held than any argumentative support we might give it, or explanatory analysis we might offer of it. It survives, sometimes in significantly qualified form, the failure to exhibit the rules which would define the rationality of science. Philosophers of science, of course, have not given up the quest to ground and explain the common wisdom. I wish to distinguish four important positions (not all of which are mutually exclusive) that are present in the current literature, all of which are worthy of further development, and the fourth of which I will attempt to develop.

The first position stresses that the appropriate rules have not yet been exhibited, and continues in the project to discover them.[13] A second position sees the quest for rules as unpromising, and instead attempts to characterize the rationality of science in terms of commitment to a set of values, *epistemic values* such as predictive accuracy, explanatory comprehensiveness, fruitfulness in investigation, internal consistency, simplicity and the like.[14] Epistemic values are sharply distinguished from pragmatic values, including moral and political values, in order to uphold the claim of the universality of scientific rationality and to deny that it is specific to the projects and interests of a particular culture.

This second approach is promising, but in need of further development. Values lack the precision and quasi-algorithmic character of rules, and thus they permit that people, acting according to the canons of rationality, may reach conflicting judgments. The lack of precision has two dimensions: people may differ in their estimates of the extent to which one or other of the epistemic values is manifested in a given theory, and in the weights they grant to the importance of each of the values. Unlike accounts of rationality in terms of rules, this seems to fit more accurately with the history of scientific practice, so much so that it appears to be plausible to portray the history of science as the history of the community of investigators committed to the set of epistemic values. However, the successful portrayal of the history of science in these terms is insufficient to establish the universal claim of these values. It would have to be shown not only that these values define the tradition of Western science, but also, that the products of this tradition make a claim on all cultures. The fact that the list of epistemic values has been arrived at through historical inquiry, rather than with *a priori* warrant, makes it difficult to establish the latter point. Another difficulty is that the weights of importance assigned to the various epistemic values may be a function of prior commitment to certain pragmatic values. E.g. the heavy weight attached to predictive accuracy in modern natural science may be due to the modern interest in technological control, and so not make a claim upon a culture which downplays this interest.

A third position sees the key feature in the rational assessment of scientific theories as residing in the recourse to *experiment*.[15] The universal claim of science rests upon its successful and demonstrable construction of replicable experiments; the claim of a theory follows from its power to explain such experiments. The credibility of science flows from the successful, premeditated interventions into the course of events which it informs. Oversimplifying considerably, one might say that the primary task of scientific theory is to explain successful experiments and those (relatively rare) natural occurrences which can be characterized with the same exactitude with which we can characterize experimental phenomena, and to anticipate further such experiments. With the task of science so specified and circumscribed, one can probably show that commitment to the epistemic values will best further its success, and with the class of empirical data limited to descriptions of such phenomena even the possibility of finding rules becomes less remote – but their role would remain secondary in the specification of scientific rationality.

Within this position one is likely to express the claims of science in a more modest light. One is less likely to speak of science as giving the best account of "the universe" or "nature" or "the physical world" or "the way things are," and more likely to speak of it as offering explanations of particular kinds of objects or of natural particulars or of objects under specified boundary conditions, and of the principles of science as expressing tendencies of nature, explicitly manifested under certain closed conditions, rather than generalizations of nature instantiated in all circumstances.[16] This claim for science, modest though it is, still upholds the universal aspiration of scientific rationality. It maintains that successful, replicable experimental intervention demonstrates the superiority of scientific theories for producing understanding of objects under closed boundary conditions, and this understanding is produced through the representations expressed in theories of the tendencies and constituents of nature at large. Rationality requires deference to such scientific knowledge in the sense that any belief or practice (in any culture) inconsistent with such knowledge is to be rejected. As I have stated this third position, however, it does not seem to require the universal further nurturing of science among rational people, since it remains open that the interest of a particular culture might not be furthered through engagement with experimental objects.

The fourth position incorporates the third but goes beyond it. Like the third, it sees the root of the superiority of scientific rationality in successful interventions in the world that have been informed by scientific theory, but it highlights practically motivated interventions, or interventions which engender technological control. Scientific understanding is superior because it engenders technological control – a move to the second perspective of the opening paragraph. Bacon held such a view at the outset of modern science, but the mainstream of philosophy of science has consciously rejected it, because the achievement of control expresses a pragmatic value, and this is seen as running contrary to the universal aspirations of scientific rationality. More problematically, it seems to involve an inference of fact from value, the inference from "theoretical science serves the desired end of informing technological control" to "theoretical science provides the best understanding of the physical world." These certainly are difficulties that must be confronted by one who would hold this particular position. In a recent article,[17] Charles Taylor has argued that the superiority of scientific understanding is linked with its informing technological control, and that it is precisely this link that accounts for the universal claims that scientific rationality makes.

IV

Taylor's argument is complex, subtle, in places even elusive, and in need of careful exposition. He concludes that what he calls *theoretical understanding* is superior in rationality to other (both earlier Western and contemporary non-Western) forms of understanding *of the physical world* because of the technological control that it informs. By "theoretical understanding" he means that form of understanding in which states of affairs and systems are explained by relating them to the principles and posits of modern scientific theories. He also refers to it as the form of understanding that derives from "the disengaged perspective." Although this is an important characterization of theoretical understanding it does not impinge significantly upon the main line of argument which I will be discussing, so I will omit comment on it.[18] I will reconstruct Taylor's argument concisely, attempting to make its principal moves clear, interspersed with a few comments aimed at clarification, then in the next sections I will offer some criticisms.

The basic structure of the argument is given in propositions 1–6.

1. A form of understanding, in order to be considered rational, must be consistent.

Consistency, while necessary, is insufficient for the full characterization of rationality, for it alone does not permit the critical comparison of forms of understanding across cultures.

2. To understand a thing ("to have a rational grasp of something")[19] is to have an *articulation* of it – to be able "to distinguish and lay out the different features of the matter in perspicuous order."
3. "The best articulation of something is what lays it out in the most perspicuous order" – so that greater rationality accrues to a form of understanding which enables a more perspicuous order to be expressed.
4. Greater perspicuity comes with "a broader, more comprehensive grasp on things."

The notions of "perspicuous order" and "broader, more comprehensive grasp on things" raise some difficulties of interpretation which I will pick up, in section VI, in the course of my critical remarks.

5. Theoretical understanding offers a broader, more comprehensive grasp on things. (I will call this the *Comprehensiveness Thesis*.)

It follows that

6. Theoretical understanding is the bearer of superior rationality.

Taylor holds that theoretical understanding is appropriately sought only for things in the physical world, not for human phenomena, but this detail need not detain us. For the remainder of the discussion, let it be understood that I am only talking about understanding physical things. Taylor himself expresses the conclusion with the words, ". . . modern science represents a superior understanding of the universe, or if you like, the physical universe . . .. [It offers] greater understanding at least of physical nature.[20]

It is clear that the Comprehensiveness Thesis, proposition 5, is the key to the argument. I construe much of Taylor's paper as an argument for it. Propositions 7–15 represent my reconstruction of the argument. (Some of the propositions I state are not actual moves in the argument, but rather represent presuppositions that make some of the premises intelligible.)

7. There are alternative forms of understanding.

E.g., in an earlier phase of Western civilization there was a form of understanding in which "understanding" and "attunement" were inseparable, and in other cultures there are forms of understanding in which our separation between practical activity and symbolic expression cannot be made.

8. To each form of understanding there corresponds a *characteristic activity* which (a) the form of understanding illuminates, and (b) contributes towards the generation or consolidation of the form of understanding.

"Characteristic activity" is my term; Taylor uses such terms as "mode of activity" and "form of life." E.g. corresponding to that form of understanding in which understanding and attunement are inseparable there is the characteristic activity of adaptation to nature, and to theoretical understanding there is that of exercising control over nature.[21]

9. The characteristic activity of theoretical understanding is incommensurable with that of earlier phases of Western civilization, and those of other cultures.[22]

Taylor has an interesting account of incommensurable activities. They "are rivals; their constitutive rules prescribe in contradiction to each other." "They are different, yet they somehow occupy the same space." They differ concerning the "range of possible activities, that is the way in which they identify and distinguish activities . . . [incommensurable activities] cut across ours in disconcerting ways."[23] In this last bit, Taylor is referring to the practice of magic and our inability to locate it adequately either within practical or symbolic activity. Looking in the opposite direction, Aristotelians had difficulty in locating Galilean science. Galileo had his Aristotelian protagonist, Simplicio, express the perplexity thus:

> these mathematical subtleties do very well in the abstract, but they do not work out when applied to sensible and physical matters . . . when it comes to matter, things happen otherwise[24].

Simplicio was unable to locate mathematical physics adequately either as applying to nature or to artifacts. The characteristic activities of adaptation and exercising control are clearly incommensurable in this sense. They are not merely different, so that one could imagine them going on together. The very success in exercising control changes the environment in which one lives, yet adaptation presupposes a more or less constant environment, subject at most to periodic rhythms; and conversely the predominance of adaptation precludes the kind of exercise of control necessary for the development of scientific theory.

At this point, an extended digression from the exposition of Taylor's argument will be useful for purposes of clarity and to forestall possible misunderstandings. Both of the characteristic activities under discussion, adaptation and control, involve interventions into nature, since – no matter what the prevailing predominant stance towards nature is – human beings do things to natural objects in order to realize certain of their ends. There is no *human* life that does not involve some form of intervention in this sense; that is characteristic of human nature. In a broad sense, all living entities intervene in nature, through various mechanisms of assimilation and accomodation. The human distinctiveness is that the intervention is purposive and planned, expressive of stances which are illuminated by forms of understanding.

Intervention, within the adaptation stance, takes place within natural rhythms, with a limited set of ends and means defined by traditional practice. The tradition ensures that, save for unexpected circumstances, there is a harmonious and reciprocal interaction between humans and nature. Human interaction is balanced by a nourishment and maintenance of the natural environment, so that human relationships with it can remain permanent. The preservation of this environment sets bounds to acceptable ends. Such a constancy, punctured only by periodic variations, can provide the basis for a stable social order, where there is deep interpenetration of social and cosmic visions. Since the stability is premised on natural constancy, such orders are exceedingly vulnerable to natural disasters and powerful outside intrusion.[25] In contrast, control is part of a stance where social order becomes subsidiary to material goals. Ends are not circumscribed by a natural environment which is to be maintained, but only by the possible, and the power to implement the possible.[26] That is why control and adaptation are incommensurable.

The sort of incommensurability discussed by Taylor is deep, deeper than the notions of "incommensurability" of Kuhn and Feyerabend where the primary reference is not to activities but propositions, theories and meanings. Suppose one lives within a culture engaged in the characteristic activity of adaptation. Its form of understanding has both empirical and moral support; it makes sense of a good deal of lived experience and outlines the path to attunement. It also delimits the class of the possible, and denies the possibility of attunement if one engages in different activity (and all of this may be incorporated into a rich theological vision.) There is a sort of self-enclosure that does not allow space for alternative activities such as the exercise of control – except on the margins – just as the modern Western enclosure, defined by the shifting perspectives of scientific knowledge and technological control leaves little space for alternative activities.

Suppose one wishes to challenge its demarcation of the range of possibilities. To make good one's challenge, one would have to engage in the alternative activity, but that threatens attunement. *Before* making good on one's claims about the possibilities, one would have to engage in the other activity, thereby possibly undermining the activity of adaptation not only for oneself but for the culture, since the natural world in which adaptation takes place could be significantly changed in the course of attempting to establish one's claims. That course of action might then be vindicated retroactively should one's activity actually realize the possibilities that the adaptation course denied. But one doesn't know in advance that one will be vindicated, and in the process of acting – hoping to be vindicated – one may destroy the environment in which attunement was possible. This suggests that for the alternative activity to get under way, there would have to be doubts about attunement. At least on the margins there would have to be embryonic grounds to suggest that the alternative practice may produce betterment for humans beings. Otherwise, the self-enclosure would seem to be unbreachable.

The promise of human betterment was a major theme in virtually all the important contributors to the scientific revolution. It has been ignored in most discussions of the superiority of modern to ancient science, being held to be of no cognitive significance. Perhaps it has no bearing on the truth value of whether a certain theory best articulates a given system. But it does bear upon whether the pursuit of theoretical understanding is desirable. In the process of the scientific revolution, the notion of "human betterment" like that of "theoretical understanding" was transformed. Taylor says that not to be attuned to nature is "to be in misery and confusion." Perhaps through this negative side there is retained a thread of contact between the old notion of attunement and the modern sense of betterment.

This ends the digression, and we return to the exposition of the main lines of Taylor's argument.

10. In every culture, there is a pre-scientific understanding which informs our "ability to make our way around in [the world] and deal with the things in it."[27] This enables the transcultural recognition of effective practices and their comparison.

Taylor grounds this assertion on a general view of "the kind of beings we are," namely "embodied and active in the world."

11. The modern achievement of technological control brings into being more effective practices.

Then, there follows in Taylor's text a series of passages that I find difficult to interpret. Compare ". . . one set of practices can pose a challenge for an incommensurable interlocutor, not indeed in the language of this interlocutor, but in terms which the interlocutor cannot ignore," and "But once a spectacular degree of technological control is achieved, it commands attention and demands explanation" with "There is an inner connection between understanding the world and achieving technological control which rightly commands everyone's attention, and doesn't just justify our practices in our own eyes."[28] I draw out of this the following moves in Taylor's argument.

12. The achievement of technological control (how and why technological devices work) *requires* explanation, in any culture which comes into contact with it.[29]

I will discuss the modalities implicit in "requires" in section V.

13a. Theoretical understanding is *the* form of understanding which explains technological control.

That is the "inner connection" between understanding and control which "rightly commands everyone's attention." Taylor also talks of the technological spin-off of scientific theorizing, of theory yielding "further and more far-reaching recipes for action,"[30] suggesting that he has in mind the alternative formulation:

13b. Theoretical understanding is *the* form of understanding which systematically increases the range of technological possibilities open to us.

Technological spin-off is obviously important, and a major part of the reason why societal institutions support scientific research, but I tend to regard it as a symptom of the truth of 13a. According to the distinghuished historian Lynn White,[31] technological spin-off of theory, though dreamed of by Bacon almost 400 years ago, became a significant phenomenon only in the last 150 years. But since at least the time of Galileo, theory has been used to explain technological devices. For the purposes of the present argument, therefore, 13a rather than 13b seems to be the salient formulation.

14. Theoretical understanding displaces other hitherto existing forms of understanding (which encompases material objects) – i.e. they cannot coexist with it. (I will call this the *Displacement Thesis*.)

Note that 14 follows from 9, 12 and 13a.

15. The capability for a form of understanding to explain technological control follows from it offering a broader, more comprehensive grasp on things.

Taylor's words are, "The superiority of modern science is that it has a very simple explanation for [the achievement of control]: that it has greatly advanced our knowledge of the material world."[32] It seems clear that Taylor intends that the best account of "the material world" be that which offers "a broader more comprehensive grasp on things."

Note that the *comprehensiveness thesis*, proposition 5, then follows from 13a and 15, so that the *displacement thesis* does not play a role in its derivation. The difference between the comprehensiveness and displacement theses is very important. Schematically, the comprehensiveness thesis says: "We reject other forms of understanding because modern science is more comprehensive," while the displacement thesis says: "We reject other forms of understanding because modern science can account for certain particulars which they can't." The displacement thesis could be true and the comprehensiveness thesis false. I will argue that this is the case, and of significance.

V

In the derivation of the displacement thesis, that the achievement of technological control *requires* explanation, proposition 12, is a crucial premise. It needs interpretation. The achievement of technological control, Taylor maintains, demands explanation. Any culture which comes into contact with it must construct an account of how it is possible. The culture's account of the words, of which its participants are trying to make sense as they live in it, must also make sense of this achievement. The widespread and rapidly accelerating achievement of technological control (not any specific technological devices) can't be dismissed as a peripheral oddity or an unintriguing commonplace; it can't be ignored. Taylor insists that the modalities in these statements have logical or rational force (as distinct from being necessities imposed by force by a conqueror upon a culture, or embraced by a culture in order to resist an alien power.)[33]

Technological achievements can't be ignored, in the first place, because they are in continuity with antecedent practices which serve similar goals, but they serve the goals more effectively. But this is only part of the story. For centuries before the scientific revolution, technological devices were gradually being invented, multiplied and introduced into European life, but they remained being conceived of as products of the skill of artisans and not also as objects crying out for explanation. Taylor seems to suggest that as technological devices become ubiquitous they pass from being merely products of skill to being also objects of explanation.

All of this suggests (and for the remainder of the paper I move beyond the exposition of Taylor's text and introduce arguments of my own) that the other part of the story might be that any culture will judge some of the achievements of technological control so desirable (because certain of the "recipes for action" developed by technology will enhance practices within it) that it will seek to explain their source, in order to find systematic principles that might facilitate their further expansion.

Or alternatively, the other part of the story might be that the achievements of technology have become so widespread, commonplace and above all central that an account of the world of lived experience that does not offer an account of them is manifestly incomplete, since a significant proportion of human experience has come to involve interaction with technological objects; indeed for that part of human experience relations with nature have come to be largely mediated by technological objects. I will develop only the alternative story proposed in this paragraph, since it seems to me to offer the more promising line of development, but I do not wish to imply that the considerations of desirability of the previous paragraph are ever wholly absent.

It is fair to say that a form of understanding that reaches predominance in a culture – one that grips the minds and imaginations of the participants of the culture, so that they easily interpret their characteristic experiences in terms of it – aspires to *a kind of comprehensiveness*, such that it gives an account of the salient objects of the culture's environment, the objects constantly present to experience and, above all, the objects with which the members of the culture habitually interact. There is the aspiration that the characteristic activity associated with the form of understanding (proposition 8) constitute the hegemonic practice of the culture, or at least be the pacesetter of its trends. A form of understanding becomes predominant in a culture only if there is a world of lived experience (either predominant or pacesetting in the culture) which the form of understanding is uniquely able to illuminate in its key features. The kind of comprehensiveness referred to here always remains as aspiration, realized to a greater or lesser degree; in practice, there always remain facets of experience which elude the forms of understanding.[34]

From the late 16th century on, technological devices and their products multiplied rapidly in Europe. They were associated with new forms of prosperity and new social forces, and their multiplication generated social change. Increasingly they became objects of attention to those in power and vying for power, and their role in daily life increased. Since they were at the cutting-edge of the movement of society, it became socially important to grasp what were the hidden and far-reching possibilities of technology. To chart such possibilities, the technological achievements had to become objects of understanding.[35] Gradually, human life in Europe came to involve – for most people, for the greater part of their daily lives – interaction with the devices and products of technology. A form of understanding that failed to make sense of the technological success (and the associated successful interventions into nature in experiment), therefore, failed with respect to the aspiration for comprehensiveness. Technological success had to be understood, or else the world of lived experience would manifestly be incompletely understood, since technology had moved from the margins to the center of life. It is in this sense that technological achievement *requires* explanation, that it cannot be ignored.

The argument just presented applies clearly only to Western culture. A fuller story is needed for the conclusion to apply also to non-Western cultures which come into contact with technological achievements, for in their case the requirement, which Taylor claims is present, exists prior to the widespread implementation of technology within the society. Still, I think that Taylor's claim is basically correct. Once there is prolonged contact with a technologically advanced culture (whether enforced or not), the horizons of possibilities of the culture necessarily change and its directions become shaped by those sectors within it (or operations upon it from outside) most in contact with the technological innovations.[36] Thence in order to comprehend the movement of the culture, it is required to understand technological achievement.

Now let us see how proposition 12, that a culture cannot ignore technological achievement, understood in the manner just developed, fits into the overall argument. Proposition 13a, as stated, is true; scientific theory and no other hitherto existing form of understanding can explain how technological control is successful, how and why technological and closed systems behave as they do under certain conditions in which they are placed.[37] The displacement thesis is also true. I noted above that it follows from 9, 12 and 13a. However, it needs fuller exposition. Its truth is manifested on a number of levels.

On the first level, scientific theory entails that there are certain possibilities which are deemed impossibilities within other forms of understanding (e.g., motion in a vacuum, sending objects from earth to other planets), and sometimes they are actualized in experiment or technology. On the second level, scientific theory produces accounts of objects or systems (e.g. the planets) which play a role in both theoretical and other forms of understanding, where its account is inconsistent with general principles formulated in the other forms of understanding.

These two levels are important, but they don't quite add up to the full force of the displacement thesis. In their light only, it's not clear to me that another form of understanding would have to be displaced. Perhaps with suitable adjustment of *ad hoc* hypotheses and *ceteris paribus* conditions it could accommodate itself to these matters. But accommodation is not possible when the third level of consideration (which draws upon proposition 9) is brought into account. The third level recognizes that the characteristic activity of scientific theory, because it has been successful, has changed the world of lived experience, so that the formerly central objects of experience (which were the paradigms with respect to the older forms of understanding) became displaced, and replaced at the center or at least at the cutting-edge by technology and its products. I.e., theoretical understanding displaces (renders obsolete, meets a necessary condition for viability which they don't) other forms of understanding because it, and at the present time only it, can account for the technological innovations that have come to be indispensable or at least unavoidable in human life.

Thus, I am maintaining that theoretical understanding came to flourish (to be upheld and to develop) while other forms of understanding had to be displaced because they could not accommodate the new and important objects of lived experience. The world of lived experience changed, and other forms of understanding could not account for the salient objects of the new order of experience. Theoretical understanding does offer an account of these salient objects. Thus, the other forms of understanding cease to be viable. Once one accepts that one lives in a world in which technological control is at center-stage, one has to discount other forms of understanding.[38]

Notoriously, recent writers in the philosophy of science have been finding it difficult to give an account of the way in which the transition from Aristotelian to Newtonian science is rational. They have sought to explicate this rationality in terms of the role of certain rules and/or values in assessing scientific claims (see sections II and III), without reference to the world of practical life, remaining thereby at the first two levels discussed above. The complete answer, I have suggested following Taylor's hints, is not to be found there. Now let us add the simple principle: "a viable theory of material objects must be able to offer an account of the material systems of greatest salience to the current productive activities, especially these activities that are the key to shaping the social order," a principle justified on the grounds that we expect of a form of understanding that it first and foremost make sense of the world of lived experience. Then, given the alternatives at hand, we have no rational alternative to being Newtonians rather than Aristotelians.[39] I think that the addition of considerations like these would significantly improve Kuhn's account of the transition to modern science.

VI

It is clear that one cannot move from the displacement thesis to the comprehensiveness thesis. Scientific theory might displace all extant alternative forms of understanding in virtue of its clear superiority with respect to explaining a special set of phenomena – experimental and technological interventions – but without having an especially comprehensive story to offer. Why is Taylor not content with this restricted superiority for scientific rationality? I think there are two reasons. In the first place, displacement as analyzed doesn't seem to be strong enough to support the claim that a rational culture must incorporate in its forms of life the further quest for theoretical understanding. It would not do so if it found technological objects to be undesirable and it could envisage a future without contact with them. In the second place, comprehensivenss does seem to be tied to our notion of rationality. For we seek to understand things and the possibilities open to us as guides to action, as informants to what we might do and make. The broader, more comprehensive articulation, the bigger picture would thus seem to be that which would better lead to a unity between our doings and makings, and the goals we set for ourselves and our culture. At this point we return to some of the notions which are key to the earlier premises of Taylor's argument (section IV), and which need further clarification.

According to Taylor, understanding involves articulation, being able to express in words distinctions, order, relations that make for perspicuity – which enables us to plan, to anticipate, to know what is a given, etc. Taylor does not elaborate, but it seems to me that *understanding a thing* (Taylor's idiom) involves being able to express such matters as the following. (I leave it open whether or not my list is complete, and the extent to which the items overlap.)

(i) The principles with reference to which we can account for its movements and variations.
(ii) What possibilities are open to it: to develop into, to affect as a causal agent, to become upon decomposition or upon becoming a constituent of other objects or systems; and the conditions under which these possibilities are realized.
(iii) What its components are, and the manner in which they are structured.
(iv) The conditions which brought it into being, and those which sustain its existence.
(v) What structure or structures it is part of, and how variations in it have impact on such structures (and conversely) – more generally, how variations in it have impact upon the environment (physical, biological, human) in which it is located, and how variations of the general environment have impact on it.
(vi) What other objects or systems it is related to in virtue of sharing the same explanatory principles – and what accounts for its variations being different from theirs.[40]

I will call a form of understanding that offers expressions, *for an extensive range of the objects of lived experience*, under all six items *comprehensive*; one that offers them of (i)-(v) *full*; and one that offers them of (i)-(iii), (vi) *wide-ranging*.

Now let us return to the main argument. We have seen that we can't pass from the displacement to the comprehensiveness thesis. The role of proposition 15 is to enable a move from 13a to the comprehensiveness thesis. There is something appealing about this move. Most of us hold, with Bacon, that mastery of nature is a consequence of having grasped the laws of nature. Indeed, theoretical science helps to generate, as well as explains, technological control because *theoretical understanding is wide-ranging*. (E.g. it relates the motion of an arrow to those of the planets and falling bodies.) But *it is not full*. (It doesn't relate the arrow to the archer, the victim, or the artisan who crafted it.) In not offering a full account, theoretical understanding not only does not treat things as cultural objects (of course, some things are not cultural objects) nor their effects on human lives, but also does not deal centrally and sometimes not at all with the side-effects of technological interventions including their environmental consequences.

Theoretical understanding deals with things solely under physicalistic description – a mode of understanding which underlies the power, *given certain boundary conditions* (which may be explicitly elaborated, or demarcated only by *ceteris paribus* clauses) to predict and control, and thence the power to uncover novel possibilities of nature. Nevertheless, its mode of dealing with *most* of the objects we encounter in daily life involves abstraction. (An arrow is an arrow, a cultural artifact, not simply a material object characterized by a certain mass, etc.) An older terminology would say that theoretical understanding deals only with the "material causes" of things. The gain is in technological advance; the problem is that technological objects are introduced without a *full* understanding, and this is manifested in current environmental and ecological crises.

There are those who think that the wide-rangingness of theoretical understanding will eventually generate fullness, that physicalistic accounts will eventually be full accounts and that all causation will be reducible to material causation; and relatedly, that every environmental problem will eventually have a technological solution. I have no strong views on these matters – only a light-hearted skepticism. I am confident that they do not have the support of *a priori* arguments, and hence arguments concerning them tend to become exercises in utopian speculation, rather than compelling guides to the quest for understanding or the practice of science.

As things now stand, theoretical understanding is wide-ranging[41] but not full. As such, it is an incomplete form of understanding for dealing with the objects of our lived experience. For this we need to be informed not only about available and novel possibilities, but also about their far-reaching effects upon the environment and upon society. Acting in the light of theoretical understanding alone is to act with a studied blindness towards such implications. That is not the mark of the rational.

In the light of this discussion, we find that the comprehensiveness thesis remains unsupported by the argument; indeed that it is false,[42] except perhaps by default; and so the argument does not support that theoretical understanding must be the bearer of a superior rationality concerning material things. But we are not left merely with the displacement thesis. I stated above that theoretical understanding explains and helps to generate technological control because it is wide-ranging. It is wide-ranging across salient objects (see section V) though by no means across all the objects of our current lived experience. More importantly for the present consideration, it ranges over a vast class of closed systems, mainly experimental constructions but also some naturally occurring. With respect to these systems theoretical principles are manifestly explanatorily sufficient, as shown by the criteria of prediction and control. Theoretical understanding is grounded in experimental interventions,[43] and insofar as it is soundly grounded it can anticipate the outcome of further experimental – and by extension – technological interventions. In this way, theoretical understanding grasps the possibilities of things, the possibilities they will realize under a wide range of boundary conditions, which are characterized (within theoretical understanding) in purely physicalistic terms. That is why I referred to it as dealing with the "material causes" of things, i.e. the possibilities that are open to things considered simply as material things (abstracted from any cultural role), and the material conditions under which these possibilities would be realized. I do not doubt that modern science provides a superior understanding of the material causes of things to that of earlier forms of understanding, simply because it has generated knowledge of far more material possibilities. In fact, I am inclined to the view that the great power of modern science derives from its discovery that material causes could be investigated and understood consequent upon experimental interventions into the normal course of things.

But is this conclusion any different from that of Taylor, who after all maintains that theoretical understanding gives a better account of "the material world"? I think it is. I resist Taylor's idiom and what I think is its intended content. Taylor refers to theoretical understanding as derived from a disengaged perspective[44] on the material world. This suggests that it expresses understanding of material objects applicable across perspectives of engagement, as they are independently of human interactions with them, and underlying their role in all cultures and epochs. It suggests that theoretical understanding is moving towards an account which would represent *all* material possibilities (at least for important classes of material objects). In contrast, I maintain that theoretical understanding derives from adopting the technological stance towards material things, from experimental (and by extension, technological) interventions into the course of things and then charting the regularities under the created boundary conditions. On this view, theoretical understanding represents the possibilities of material things under a range of boundary conditons.[45]

My view is sufficient to account for technological success, and it is unencumbered by the gratuitous universality of the former view.[46] Moreover, on my view one will be less tempted to identify wide-ranging with comprehensive. For, in many cases, the obtaining of the boundary conditions (though for purposes of scientific theory they are characterized in physicalistic terms) is not itself explicable in physicalistic terms, but only in terms of human parameters and social conditions.[47] As such, given different social boundary conditions the object of inquiry *may* differ.

VII

Taylor concluded, drawing upon the intimate connection between science and technology, that theoretical understanding of material objects is the bearer of superior rationality. I have shown that his argument fails. However, from the same starting-point weaker, though still important conclusions follow, namely that theoretical understanding displaces other hitherto existing forms of understanding, and that it is superior for dealing with the material causes of things. But it lacks the power to provide a full understanding of things.

At the outset of the paper, I expressed a common wisdom that commitment to rationality requires a way of life informed by, or at least aspiring to be informed by the claims of the currently best established scientific views, and a culture which fosters the pursuit of science bears a superior rationality. The lack of fullness of theoretical understanding bears upon this common wisdom, and so does the particular manifestation of it in the social boundary conditions of many objects of inquiry. It is insufficient for being rational that a way of life be informed by the best available scientific theories, if this be despite the social conditions and implications of its being so informed. Equally, it is unnecessary for being rational that a way of life embrace concretely scientific applications that involve social conditions which it deems undesirable. But the positive conclusion of the previous paragraph stands. It does not follow from this that any culture, in the name of rationality, need embrace the unbridled pursuit of a science that marks our culture. Given different social conditions, the objects of inquiry and the social modes of investigation *may* differ.

NOTES

1 The writing of this paper was supported in part by a grant from the National Science Foundation (SES-8308604). Earlier versions were presented at the University of Southern California, the Mid-States Philosophy Meeting, and at several working sessions of the Greater Philadelphia Philosophy Consortium. Comments by Geoffrey Joseph and Richard Eldridge have been especially helpful.

2 Even politically, relativism is a two-edged sword. While, for some, it is the foundation for recognizing the legitimacy of *difference* among human societies and practices, for others it justifies opting out of constructive interaction with different cultures since, they point out, if standards of rationality are culture-bound then rational discourse can't settle disagreements across cultures. The first outlook confuses human variety with relativism. Human variety requires only that there be significantly different conditions within which humans have to construct their lives; it does not rest upon different standards of rationality. The second outlook provides a ready rationale to settle fundamental conflicts by force, rather than a constructive dialogue that can hope for an eventual synthesis that can demand consensus. Note that if variety does spring from different conditions of life then the very act of dialogue can change the conditions. Non-relativism does not require the obliteration of variety. Underlying the argument of this paper is the non-relativistic stance that recognizes both the appropriateness of human variety and the possibility of arriving at synthesis across variety.

3 It is symptomatic rather than constitutive of a soundly established scientific theory, the predominant view maintains, that it issue in technological applications. This is compatible with technological successes constituting *part* of the evidence for a theory, and with technological devices being indispensible for the collection of relevant evidence.

4 A criterion could be both a reflection of a given culture's interests and universally valid. Thus, to deny the universal validity of a criterion, it is not sufficient to display its historical origin in a contingent practice of a particular culture.
5 Neo-Cartesian rationalism falls under this consensus, though it adds the further requirement that the class of acceptable theoretical hypotheses is constrained by consistency with *a priori* derived principles. Such constraints severely limit the class of theoretical candidates; the logical relations with data then enable the selection from among this limited class. See H. Lacey, "Constraints upon Theories," *Pacific Philosophical Quarterly* (forthcoming).
6 L.J. Cohen, *The Implications of Induction*, London: Methuen, 1970.
7 W.C. Salmon, *The Foundations of Scientific Inference*, Pittsburgh: University of Pittsburgh Press, 1967.
8 K. Popper, *The Logic of Scientific Discovery*, New York: Harper and Row, 1959; B. van Fraassen, *The Scientific Image*, Oxford: Clarendon Press, 1980.
9 C. Glymour's *Theory and Evidence*, Princeton: Princeton University Press, 1980.
10 I. Lakatos, *The Methodology of Scientific Research Programmes, Philosophical Papers*, vol. 1, Cambridge: Cambridge University Press, 1980.
11 An important example of this occurs when Popper (*op. cit.*, p. 103) introduces conventions to identify the class of "basic statements." Basic statements are about observable events, where an observable event is "an event involving position and momentum of macroscopic physical bodies." Limiting the class of empirical data in this way is sufficient to make a theory like Aristotle's, which builds upon qualitative categories, a non-competitor.
12 The key texts are T. Kuhn, *The Structure of Scientific Revolutions*, Chicago: The University of Chicago Press, 1962, and P. Feyerabend, *Against Method*, London: New Left Books, 1975. R. Bernstein, *Beyond Objectivism and Relativism*, Philadelphia: University of Philadelphia Press, 1983, offers a balanced assessment of the contribution of these two authors.
13 W.C. Salmon, *op cit.*, and "Carl C. Hempel on the Rationality of Science," *The Journal of Philosophy 80* (1983), 555–563.
14 T. Kuhn, "Objectivity, Value Judgment and Theory Choice," *The Essential Tension*, Chicago: The University of Chicago Press, 1977; E. McMullan, "Values in Science" in P.D. Asquith and T. Nickles (eds.), *PSA 1982*, vol. 2, East Lansing: Philosophy of Science Association, 1983, pp. 1–25. C. Hempel ("Values and Objectivity in Science," in R.S. Cohen and L. Laudan (eds.), *Physics, Philosophy and Psychoanalysis*, Dordrecht: Reidel, 1983, pp. 73–100) now finds this approach congenial at least as a temporary expedient, though he has not quite given up on the long-term quest for rules.
15 The special significance of experiment in scientific inference was strangely not emphasized in the writings of the logical empiricists and their early critics. They stressed the observation of nature, rather than intervention in it. Among the recent writings which have put the central emphasis upon experiment are R. Bhaskar, *A Realist Theory of Science*, Atlantic Highlands: Humanities Press, 1975; I. Hacking, *Representation and Intervention*, Cambridge: Cambridge University Press, 1983; N. Cartwright, *How the Laws of Physics Lie*, Oxford: Clarendon Press, 1983.
16 These ways of speaking derive mainly from Bhaskar (see previous footnote). The influence of Bhaskar is pervasive in this paper. I acknowledge it here rather than attempt to indicate every instance of the influence. Hacking and Cartwright also tend to deny the truth or widespread applicability of general theories, and speak instead of entities that have been discovered as the causal structures of experimental phenomena, have been uncovered with the aid of theoretically informed interventions. All three resist the ready move from the demonstrated adequacy of a theory to explain experimental phenomena (and other phenomena in closed systems) to the assertion that the theory represents significant or dominant explanatory features of nature at large.
17 C. Taylor, "Rationality" in M. Hollis and S. Lukes (eds.), *Rationality and Relativism*, Cambridge: HIT Press, 1982, pp. 87–105.
18 In a sequal to this paper I expect to take up discussion of the notion of "the disengaged perspective." Let me here only record my conviction that theoretical understanding derives not from a disengaged perspective but from adopting what I call "the technological stance" towards material things.

19 All quotations in 1–6 are from Taylor, p. 90.
20 *Ibid.*, pp. 102–103.
21 It is easy to oversimplify here. The characteristic activities reflect what I would call predominant stances towards nature. They are mutually exclusive as predominant stances, but in the lived experience of a culture, elements of both are always present. The way a culture seeks understanding reflects the predominant stance, rather than activities engaged in at its margins. However, marginal activity can be the source of new forms of understanding.
22 Sometimes Taylor speaks of the activities of understanding (or gaining understanding) as incommensurable, but these activities are embedded in the respective characteristic activities.
23 *Ibid.*, pp. 98–99.
24 Galileo, *Dialogue Concerning the Two Chief Systems of the World*, Berkeley, University of California Press, 1962.
25 Conversation with Rosemary Desjardins helped to improve the formulation of the previous few paragraphs.
26 This way of putting things suggests a question which is implicit in the discussion at the end of section VI: Can there be material progress that is real, and thence requiring social change, which is guided by social variables, such that only those technologies are implemented which will be part of the material base of a specified social structure? An affirmative answer implies the possibility of a new characteristic activity which would be a creative synthesis of the stances of adaptation and control.
27 Taylor, *op. cit.* p. 101.
28 *Ibid.*, pp. 103, 101.
29 From the context it seems clear that it is the material causation of technological objects that Taylor thinks *requires* explanation. I don't know whether or not he thinks there is the same requirement with respect to their social causation and their social effects. Note that Taylor maintains that proposition 12 represents a transcultural criterion, but one that came into being only once widespread technological control became a reality.
30 *Ibid.*, p. 101.
31 L. White, Jr., "The Historical Roots of Our Ecologic Crisis," *Machina ex Deo*, Cambridge, M.I.T. Press, 1968.
32 Taylor, *op. cit.*, p. 103.
33 *Ibid.*, p. 104.
34 These facets change with time, and are especially important at those historical moments when events, often externally generated, extend the boundaries of the environment in which the culture lives. Marginal facets can become the point from which a massive change in a cultural environment ripples out – whether it be from within, such as the developments of artisans in late medieval Europe or the invasion of foreigners in non-European cultures.
35 Galileo testified concerning the indispensibility to his theoretical and experimental inquiries of the machines and instruments constructed by skilled artisans in the Venetian arsenal. Galileo, *Dialogues Concerning Two New Sciences*, New York: MacMillan, 1914, p. 1.
36 The mechanisms of this shaping can be complex and various. Colonial force is a common factor, but Taylor is right to emphasize the continuity of technological with traditional practices, so that members of a culture can embrace a technological practice because it performs a traditional task more efficaciously. My suggestion is that Taylor's requirement holds when the culture's contact with technology has been sufficiently prolonged that technological activity has become either predominant or pace-setting in the life of the society. It does not seem to be possible in the contemporary world for a culture to isolate itself significantly from the encroachments of technological advance, so much so that technological activity has become a major factor in shaping the future of every culture, even if only to the extent of threatening to destroy the space within which a particular culture thrived. So, the requirement to comprehend technological achievement does seem to apply universally to cultures.
37 Scientific theory does not explain the social conditions required for technological success – the commitment of resources to research and development, the order of the workplace, the relevant class structure, the international order needed for the supply of materials, etc. It explains, once given these conditions, how the machines (and related systems) run, and how

other interventions work. It also does not explain the human and social consequences of widespread technological practice. Scientific theory develops in independence of explicit considerations of attunement (or any of its successor notions). From this separation, Taylor maintains, derives the explanatory power of modern science and its capability for making novel interventions in the course of things.

38 I want to reiterate that the other forms of understanding referred to above are those which pre-date theoretical understanding, and that theoretical understanding displaces them as an account of material objects. The following are not precluded by this displacement. (i) There are (or will come into being) alternative forms of understanding in which understanding of material objects is not separable from human and social considerations, in which theoretical understanding is contained within a mere comprehensive social vision. Consistency with the scientific account of material objects (and not consistency with any materialist world-view or methaphysics of physicalist reductionism) is required of any such form of understanding. (ii) Such alternative, viable forms of understanding may be continuous with older forms of understanding: they may result from the development and radical redeployment of older forms in the context of modern lived experience. Wisdom, derived from older forms, may yet come to illuminate the modern order. (iii) There might be important residue to rescue from the older forms concerning particular material objects and particular practices. Indeed, one would expect this because older practices presumably survived because they were soundly based empirically. (See, e.g., R. Ambrose, "Agricultural Research and Breaking the Cycle of Dependency," *Science for the People 15* (1983), No. 6.), 6–31.) A particular practice is not necessarily better than an incommensurate rival because it is technological, and the other traditional. Thus, the displacement need not be a wholesale obliteration; it can involve more or fewer continuities with the past.

39 On this view, it is misleading to refer to the scientific revolution of the 17th century as "the Galilean Revolution," where the central Galilean text is the *Two New Sciences* rather than *The Two Chief World Systems*. The new account of the planetary system is, of course, a spectacular achievement and of enormous importance since it demonstrated that adopting the technological stance could illuminate naturally occurring systems as well as the spaces within which human control is exercised. Still, the rational dynamic of modern science depends primarily upon experimental and technological successes.

40 This formulation derives from C. Taylor, "The Explanation of Purposive Behavior," in R. Borger and F. Cioffi (eds.), *Explanation in the Behavioral Science*, Cambridge: Cambridge University Press, 1970, pp. 49–79.

41 Wide-rangingness is an important internal criterion (representing one of the central epistemic values) of the enterprise of modern science. *Ceteris paribus*, within modern science, a theory is preferred to a rival if its explanatory scope is more wide-ranging. It does not follow that the same criterion suffices to select theoretical understanding as superior, in an unqualified way, to earlier forms of understanding.

42 My argument is that theoretical understanding fails with respect to a major parameter of comprehensiveness, with respect to which some of its earlier competitors were strong while they, of course, were weak with respect to wide-rangingness. Given these different strengths and weaknesses one cannot affirm that, e.g. modern science offers a more comprehensive account than Aristotelian science.

43 That is true also when it applies to non-experimental objects, or objects in non-experimental conditions. I will develop this point in the sequel to this paper mentioned in f.n. 18.

44 See f.n. 18.

45 A loose, incomplete way of putting the point would be: theory represents what things can do under conditions that we can bring about, rather than what things can do under all conditions.

46 Several recent writers (e.g. Bernstein, *op.cit.*) have tried to exorcise the grip which "the Cartesian anxiety" has upon theories of rationality. I think that neo-Cartesian metaphysics has an even greater grip upon theories of the rationality of science. The notion of "the material world" comes from this metaphysics, the notion that the totality of material entities constituted a world that is self-contained, in which the entities are (in principle) fully characterizable in physicalistic terms and fully explicable in terms of laws that involve only physicalistic parameters. Exactly how the material world so conceived relates with the objects

and events of lived experience constitutes a problem that has never been solved. On this metaphysical view, the closed systems, which are well articulated by theoretical science, are simple, less complex instances of the way all natural systems are, so that the terms used in these articulations, added to and appropriately complexified, are the appropriate terms for proper articulations of the material world. There is an alternative metaphysical view which I call neo-Baconian (because of Bacon's repeated opposition to universalist aspirations) according to which the closed systems provide us with clear manifestations of the possibilities that can be realized under given closure conditions, it being open whether or not the possibilities are only limiting cases, exhibited under those conditions but not extrapolable outside of them. On this view, the physicalistic account is the appropriate account to give of entities for the sake of the objective of intervention and control (and of entities which occur naturally under conditions analogous to those under which successful control occurs). On both views it is proper to engage in research to extend the range of applicability of theoretical principles – on the former view one will thereby be filling in more of the missing pieces of the material world; on the latter one may be generating novel possibilities. On the former view, the world is a vast spatio-temporal continuum whose regions are to be charted in fine detail; on the latter it is a vast repository of possibilities to be plumbed in various ways to see what it will realize. The outlook expressed in the main body of the paper is influenced by the neo-Baconian metaphysical view, but not necessarily committed to it. I highlight it mainly to show that scientific practice is not necessarily wedded to the neo-Cartesian view. Both views are compatible with the practice of science. See H. Lacey, "Constraints Upon Theories," *op.cit.*

47 This does not relativize the truth of scientific claims to social conditions. Scientific theory explains such things as: given certain boundary conditions such and such a regularity obtains. It is irrelevant to the truth value of this claim that the boundary conditions are explained in terms of social conditions. The fact of the social conditions, however, is relevant to the interest of the claim and to the value of the research program of which it is a part.

HEURISTICS FOR SCIENTIFIC AND LITERARY CREATIVITY: THE ROLE OF MODELS, ANALOGIES, AND METAPHORS*

EUGENE LASHCHYK

1. Origins of the idea of creation and creativity

The concept of creation in Western thought has meant the act of bringing into being what was previously part of non-being. This concept can be traced at least to the medieval notion of God, the creator, bringing into existence our universe *ex nihilo*. Man was not associated with the process of creation except possibly in the arts. But even in the arts and literature, at least until the Renaissance, the artist and writer were viewed not as individuals who bring into existence forms that previously did not exist, but as persons who replicate forms found in nature. Representational art was precisely what the word signifies, a picturing of something that was already there to begin with. Likewise, in Greek literature the dominant idea was that the Greek dramatists recreated stories that were part of their cultural heritage. Aristotle called such recreation *mimesis*, which meant that tragedy was an imitation of action, an imitation of actual life.

The concept of discovery, on the other hand, brings to mind the uncovering of something that was hidden from view but was part of the real world all the time. Many methods were proposed by philosophers and scientists for uncovering the real. In spite of the fact that classical epistemological theories did not utilize the concept of either creation or discovery, they were all candidates for a type of logic of discovery.

2. Why did traditional epistemologies conspire to keep the problem of creativity submerged for so long?

The story of how the concept of creativity gained a foothold in philosophy and science is a complex one. For the most part until the twentieth century there was little or no room for the idea of "creativity" in philosophy or science. In

Margolis, J., Krausz, M. and Burian, R.M., (eds.), Rationality, Relativism and the Human Sciences.
 ISBN 90-247-3271-9.

philosophy epistemological theories promised indubitable knowledge of being, at least in metaphysics in such philosophies as Platonism, Aristotelianism, Thomism, Cartesianism, Hegelianism, Marxism, Phenomenology, and others.[1] There are exceptions to the above philosophies, but they are not part of the mainstream of philosophy.[2] In these theories knowledge acquisition was reduced to a method with the result that the concept of creativity was not needed as a category of explanation. Discussion of creativity could be damaged or completely squashed either by reducing methodology to a mechanical procedure or by insisting that one has or should have demonstrable proof when it comes to the question of justification (i.e. Galileo).

For purposes of illustration I will highlight only a few of these epistemologies. For example, concerning the medieval philosophies it can be said that God was the creator, man was a contemplator. The goal of man was contemplation of the eternal verities and the pursuit of wisdom. Even happiness was defined by Aristotle and St. Thomas Aquinas in those terms. The task of the knower was to reflect reality rather than to create anything in the process of knowing. For example, in the Aristotelian-Thomistic epistemology knowledge of the world was guaranteed by the powers of the mind, specifically the powers of abstraction.

Enough high-powered machinery was packed into the head to guarantee indubitable knowledge of the world. Universal concepts presumably derived from individuals by the process of abstraction were claimed to reflect accurately the nature or essence of individuals. The mind has the power to drop the accidental properties of individuals and concentrate on the characteristics common to all the members of certain species (e.g., humans: rationality and animality). Creativity, if discussed at all in such a framework, was relegated to the realms of music, art, and literature. All one had to do to arrive at knowledge of the world was to follow carefully the procedures of abstraction. The problems with the theory of abstraction are well known. Because there are similarities between any two randomly selected objects, the main problem for any theory of abstraction is to explain how one is to choose those similarities which define a set of objects as members of a certain natural kind.

There is another reason why creativity and science do not intersect for Aristotle. For him "scientific" knowledge was a purely deductive discipline governed by the rules of valid syllogisms.[3] Whatever ingenuity or creativity was necessary went mostly into the requirement of reducing all valid syllogisms to the first figure. I do not want to deny that deductive proof at times requires creativity. However, the type of creativity called for does not pose the central and deepest problems of creativity.

The topic of creativity in science could have surfaced with the rise of modern science but it was blocked, by Newton of all people, who insisted that he derived the laws of physics from nature.[4] The problem is that the concept of derivation as I mentioned above does not allow for the most interesting questions of creativity to surface. Derivation as normally conceived is a matter of following rules of valid inference from higher principles. Newton's insistence that he derived the laws of physics from nature has puzzled philosophers and scientists alike. The puzzle arises when one realizes that nature does not consist of propositions or principles and hence derivation in the logical sense is not possible.

Inductivist methodologies of science that began with Francis Bacon were an improvement over the Aristotelian or the later Newtonian accounts of how original hypotheses and theories were created. Still the inductivist accounts of science were a mixed blessing. Overall they probably had a stifling effect on questions of creativity or discovery in science. The emphasis on fact collecting as a preliminary starting point of scientific investigation was a misplaced task since random fact collecting cannot lead to fruitful inductions. As Hempel[5] and othere have pointed out, what the inductive method lacks is a criterion of relevance. Which facts should be collected? Certainly not all facts are relevant. Not even the formulation of the problem provides an adequate guide for fact gathering. Furthermore, the classification of facts presupposes some system of categories, a point which the inductivist tends to hide rather than illuminate. Finally, even if the above problems are solved the payoff is rather limited. True, enumerative induction can account for the creation of low-level generalizations, or, as they are often called, empirical laws of science.[6] Unfortunately inductivists are imperialists who want to generalize induction as an explanation of all science. Here the position usually called "inductivism" runs into serious problems. I have yet to see an explanation, on inductivist grounds, of Newtonian physics, or of the Einsteinian theories of relativity. An imaginative defense of a form of inductivism can be found in Mary Hesse's book *The Structure of Scientific Inference*. I am, however, skeptical that induction can explain the appearance of such original theories as Einstein's general theory of relativity.[7]

The empiricist climate was not very conducive to raising questions of creativity either in science or in literature. Part of the problem was the view, as in John Locke, that all ideas were derived either from sensation or reflection (upon the activites of the mind) or were complex ideas. Such a framework restricted the role of imagination in the formation of novel theories, which had far-reaching implications for prediction and explanation. Furthermore, in the realm of literature such a framework did not allow for an adequate account of metaphor.[8]

The major problem with the empiricists is the much too stringent claim that all *bona fide* ideas must originate in sensation. Couple this requirement with the view that the mind is passive in sensation and there is little room, if any, for the creative imagination in the generation of novel ideas.

With the rise of neo-positivism in the first quarter of the twentieth century, questions of creativity and discovery fared no better than with the classical empiricist position. The dominating theme of this period was the distinction (introduced by Hans Reichenbach) between the context of discovery and the context of justification.[9] Within this context questions of discovery in science were relegated to psychology, while questions of theory justification belonged to philosophy. Because discovery and creativity are closely related questions, creativity in science on this view was also relegated to psychology. I will argue for bringing back into philosophy problems of creativity and discovery.

An equally formidable stumbling block to the formulation of the problem of creativity in science can be found among some of the early Marxist-Leninists. I have in mind here primarily Lenin's famous epistemological treatise *Materialism and Empirio-Criticism*.[10] Ideas, in Lenin's view, are mirror images of objects in the world. When they reflect accurately objects in the outside world, they provide us with windows on the world. There is an inescapable causal link between the world and ideas in our consciousness. Such a view has the advantage that it can explain the realist beliefs held by many scientists, but at what price?

It is now commonplace to see the history of science as a sequence of incompatible theories. If one couples this view with the observation that subsequent theories drop some concepts and adopt others (e.g., phlogiston-oxygen), then what sense is there to the view that concepts reflect the world? One could say, following Rorty's ironic metaphor in *Philosophy and the Mirror of Nature*, that the mirror has to be polished further in order to reflect better. Unfortunately for Lenin's theory such a reationalization quickly loses its force with repeated use. In summary it can be said that Lenin's view on concept formation seriously underestimates the role of the creative imagination in the formulation of novel concepts and theories.

3. Creativity and Einstein's philosophy[11]

Einstein was one of the first thinkers to have developed a philosophy especially conducive to discussions of creativity in science. He rejected pure versions of both empiricism and rationalism. His philosophy comes closest to a kind of Kantianism, without the *a priori* element.

a. *Einstein's model of theory construction*

A hint of Einstein's epistemological position may be found in a letter Einstein wrote to his longtime friend Maurice Solovine.[12] There, Einstein provides a useful scheme which represents the knowledge-seeking enterprise (see Figure 1).

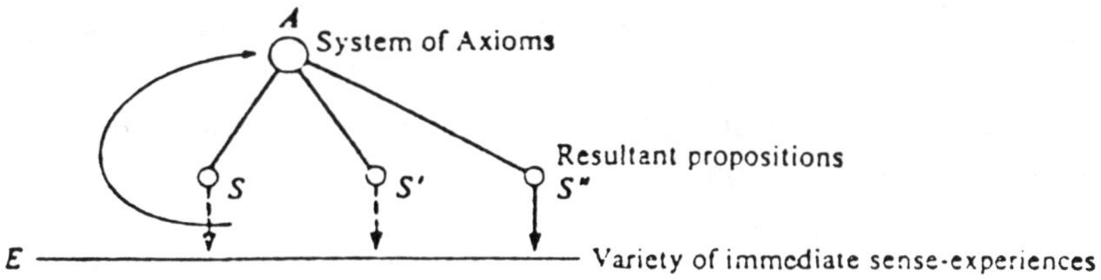

If one were to begin from the psychological perspective of the scientist engaged in theory construction, then one should start with the plethora of human experiences represented by the base line marked "E." It is clear from Einstein's writings that by this experiential base he does not mean uninterpreted givens but the whole richness of our experiences, including those which are acquired as a result of using sophisticated instrumentation (e.g., the observations that were made of the bending of star light during an eclipse of the sun). From such a potentially infinite set of experiences the scientist makes a bold creative leap, represented by the upward arrow, which I will call "J," to the axiomatic assumptions, which Einstein labels "A." There is no logical method for arriving at these axioms. According to Einstein "the scientist has to worm these general principles out of nature,"[13] but in another place he says, "only intuition, supported by being sympathetically in touch with experience" can reach them.[14]

From these general axioms, statements are logically deduced (here called "S") which are then compared with the experiential base "E." In this model Einstein speaks of concepts only at the level of assertions which are deductive consequences of the axioms. Also he feels that the relationship between the concepts in the assertion and the sensory experiences is "not of a logical nature." What is the force of this denial? He does say that as a matter of fact this relationship between the concept and the direct experience is pragmatically certain. Also he emphasizes the fact that the relation between a concept in "S" and "E" is much more certain than the far more removed relationship between the axioms and the experiential base. For fields like theology Einstein claims that since there is no contact with experience they "would be of no value at all for the 'comprehension of reality'."[15] Einstein's intention here is to underline clearly the importance of the empirical-experimental component in attempts to comprehend reality. He seems to require that any claims of knowledge about the world have an experiential connection.

But this does not explain what the relationship between the S's and the experiential base is. Mary Hesse has suggested that this relationship is metaphoric in the case of theoretical explanations.[16] Others have gone so far as to suggest that all of language applies to the world metaphorically. No matter how persuasive this latter suggestion is, the problem is that it obliterates the worthwhile distinction between literal and metaphoric meaning. I am all for keeping the distinction around.

Einstein has more to say about the whole process of concept formation even though he had little to say about the relationship between E and S. Concerning concept formation he states, in his intellectual autobiography, that "All concepts, even those which are closest to experience, are from the point of view of logic freely chosen conventions."[17] According to Einstein, concepts have a fictitious character. They are constructs and inventions rather than discoveries. But, if concepts are freely chosen conventions, if they have a conventional character, then indeed there is no logical deductive or even inductive path from the sensory experience to the concept. Actually Einstein wants to go even further. Not only are the concepts not derivable by deductive or inductive methods from experience, but neither are the laws and axioms of theoretical systems. Einstein wrote:

> A clear recognition of the erroneousness of this notion came only with the general theory of relativity which showed that one could take account of a wider range of empirical facts, and that too in a more satisfactory and complete manner, on a foundation *quite different from the Newtonian*. But quite apart from the question of the superiority of one or the other, the fictitious character of fundamental principles is perfectly evident from the fact that we can point to two essentially different principles, both of which correspond with experience to a large extent; this proves at the same time that every attempt at a logical deduction of the basic concepts and postulates of mechanics from elementary experience is doomed to failure.[18]

Einstein gives us here an important insight into the whole process of creativity in science. Implicit in Einstein's model of theory-creation is the fictitious character of the fundamental principles of any deep theory about nature. Furthermore, since it is possible to have more than one fundamental principle or theory compatible more or less with experience, the problem of theory choice is raised. For if there is no deductive or inductive method for hypothesis or theory-creation and if fundamental principles have a fictitious character, then we have the problem of how to choose among a group of competing theories that which is epistemically better. For a deductivist or an inductivist there is no problem with the justification of the chosen theory, since the logic of theory-generation functions as well for justification. Their answer is simply that their principles were derived from the data.

b. *Some extentions and problems with Einstein's model*

It should be apparent by now that the process of going from E by means of J to A, down to S and thence to a comparison with E, guarantees neither truth nor a final account of reality. It is to Einstein's credit that he realizes that there could be alternative A's with the same or partially overlapping S's, each supported by experimental data.

One such situation is presented by the Tychonic and Copernican theories, Up to the seventeenth century, both theories were almost equally supported by the same observational data. The clinching observation in favor of the heliocentric theory, the observation of parallax, was not made until 1828. But by this time most of the scientific community had been persuaded to switch to Copernicanism. Obviously more was at work than simple agreement with the data.

Schematically this extention of Einstein's original model might be represented as follows:[19]

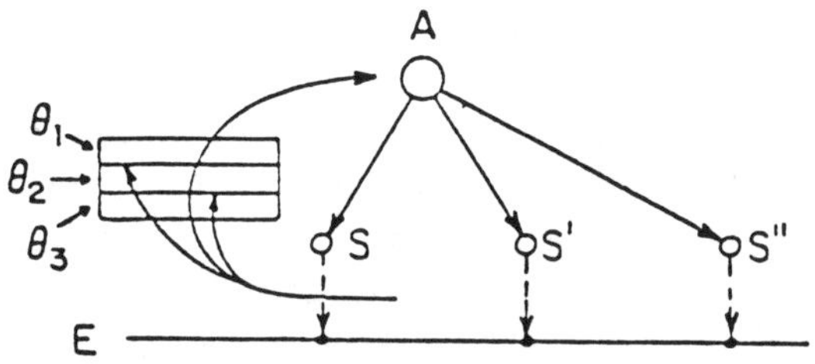

In order to explicate adequately Einstein's position on theory selection one must introduce the aim or goal of science. Only if we are clear on the goals of scientific theories can we hope to evaluate theories adequately. The EJAS model misleadingly suggests that the sole goal of science is to create theories that get confirmed in experience. It is true that the connection with experience, with experiment, is a component of good theory in modern science but for Einstein it is just one of the components of good scientific theory. For several reasons it cannot be the only requirement. In the first place, the class of theories that can satisfy this requirement is unlimited,[20] and in the second place this is not the goal of science. The goal is the comprehension of reality, not just prediction of events. Let me quote Einstein at length on this important point:

> It is, of course, universally agreed that science has to establish connections between the facts of experience, of such a kind that we can predict further occurrences from those already experienced. Indeed, according to the opinion of many positivists the completest possible accomplishment of this task is the only end of science. I do not believe, however, that so elementary

> an ideal could do much to kindle the investigator's passion, from which really great achievements have arisen. Behind the tireless efforts of the investigator there lurks a stronger, more mysterious drive: it is existence and reality that one wishes to comprehend . . .. When we strip [this] statement of its mystical elements we mean that we are seeking for the simplest possible system of thought which will bind together the observed facts . . .. The special aim which I have constantly kept before me is logical unification in the field of physics.[21]

In this passage Einstein hits upon one of the central themes of creativity in science, namely, the unification or binding together of observed facts, of experiences. This tinding together of phenomena which, prior to the binding appeared as disparate, as belonging to different categories, is a mark of a promising theory in science – provided that it satisfies other inner and outer perfections of theories. The binding together of objects from different categories is also the stuff out of which a type of metaphor is created in poetry, what Max Black has called the interaction view of metaphors.[22] It is possible to describe the most original creations in both science and poetry by using such models and metaphors.

Actually Einstein's model of theory creation presented in the letter to Solovine quoted earlier omits an important part of the enterprise of science. It provides no role for scientific problems, even though the problem gives more often than not the impetus for creating theories. The importance of problems for understanding the scientific enterprise has been stressed by many philosophers and scientists. Larry Laudan[23] is the most recent defender of the important role that problems play in our understanding of science. Laudan distinguishes at least two kinds of problems – empirical and theoretical. If the problem is empirical, as in the case of an anomalous observation (e.g., Roentgen and exposed photographic plates; Lavoisier and the gain in weight by an element which underwent combustion, under the phlogiston theory), then the problem arises in the data base "E." If the problem is of a theoretical nature, then it would be located at the top of Einstein's model.

Besides saying that the scientist has to make a creative leap (represented by "J") to the theory, we need to add to Einstein's model another element of which he speaks in other contexts. I have in mind here what Holton has called "themata," or what Einstein termed "inner perfections of a theory."[24] These inner perfections function as a filtering mechanism on the outputs of the creative imagination which generates the hypotheses and theories.

Holton is inclined to put these themata (Θ) as a screening mechanism for the proposed hypotheses and theories in the middle of the upward "J."

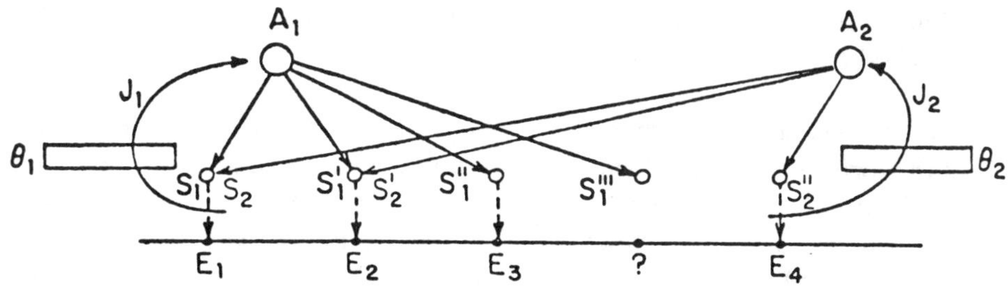

I do not want to deny this as a possible interpretation and extension of Einstein's model, based on Einstein's practice as a scientist. But I do think that in Einstein's philosophical writings he also wants to use these criteria as a filtering mechanism which strains out the less adequate of competing theories. That Einstein wants to look at these criteria more in the context of theory-selection and evaluation than in the context of theory-creation is clear from his autobiographical notes.

> The second point of view is not concerned with the relation to the material of observation but with the premises of the theory itself, with what may briefly but vaguely be characterized as the "naturalness" or "logical simplicity" of the premises (of the basic concepts and of the relations between these which are taken as a basis). This point of view, an exact formulation of which meets with great difficulties, has played an *important role in the selection and evaluations of theories since time immemorial* . . .. Furthermore, among theories of equally "simple" foundation that one is to be taken as superior which most sharply delimits the qualities of systems in the abstract (i.e., contains the most definite claims) . . .. The following I reckon as also belonging to the "inner perfection" of a theory: We prize a theory more highly if, from the logical standpoint, it is not the result of an arbitrary choice among theories which, among themselves, are of equal value and analogously constructed.[25]

The model that Einstein has in mind in the above passage would put these inner perfections of a theory at the top.

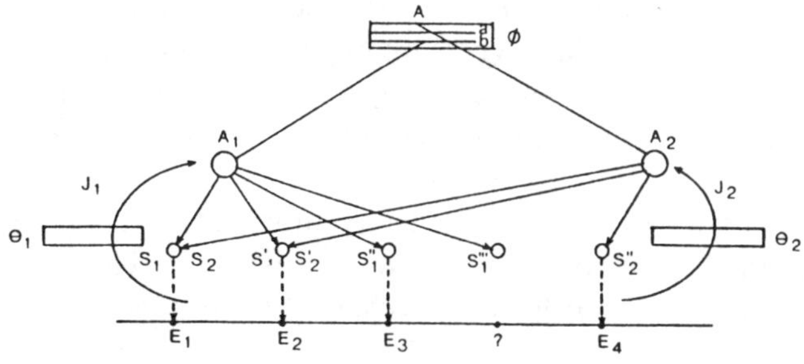

Holton, in his paper entitled "Constructing a Theory: Einstein's Model," provides the following summary of his discussion:

> Among themata which guided Einstein in theory construction are clearly these: primacy of formal (rather than materialistic) explanation; unity (or unification); cosmological scale (generalizability, and egalitarian applicability of laws throughout the total realm of experience); logical parimony and necessity; symmetry; simplicity; causality; completeness; and of course constancy and invariance. It is adherence to themata such as these which explains in specific cases why he would obstinately continue his work in a given direction even when testing against experience was difficult or unavailable.[26]

Implicit in Einstein's discussion of these themata or criteria was a certain strategy as well as a certain ranking of these criteria. Unfortunately Holton does not try to reconstruct the model of rationality implicit in Einstein's reasoning.

In a series of recent papers Thomas Nickles[27] argues for a similar role of a filtering mechanism for proposed solutions to problems, which he calls "constraints". He states "the constraints, in thus rstricting inquiry, normally provide a heuristic guide or directive. In short, a problem itself more or less points the way to its solution."

Despite the fact that Einstein's philosophical position is conducive to discussions of creativity in science, he still says precious little about the creative process itself nor does he indicate how the creative process belongs within epistemology. Einstein, nevertheless, has to be given credit for anticipating, by at least twenty years, the revolutionary philosophies of Thomas Kuhn and Paul Feyerabend, which are also so amenable to discussions of creativity and discovery. In Einstein's writings we can find many of the theses which later shocked the philosophical community in the sixties, such as the theses that observations are theory-laden; that there is no method for arriving at imaginative theories, that theoretical pluralism is a fixture of the scientific enterprise.[28]

Many reasons can be given for the fact that philosophers and scientists have begun to re-examine traditional epistemologies and have either given up the idea of epistemology or are actively searching for new alternatives. High on the list of reasons for a crisis in epistemology are the following: (1) Studies in the history of science have revealed that science developed not by simple accretion but by revolutions, discontinuous quantum leaps.[29] (2) The development of alternative systems of geometry in the 19th century raised the questions: Which of them, if any, correctly describe our world?[30] (3) The development of alternative systems of n-valued logics have raised the same question for reasoning. (4) Studies in cultural anthropology have raised questions of relativism not only in morals but also in reasoning, metaphysics, and science.[31] (5) Theories are underdetermined by the data, and empirical confirmation is not sufficient for truth.[32]

Because of these and other reasons, metaphors for gaining knowledge, such as uncovering or mirroring,[33] are under attack. Previous accounts of logic of discovery which guaranteed indubitable knowledge have to be scrapped.

4. Thomas Kuhn, scientific revolutions, and creativity

In spite of the fact that Einstein anticipated many of the outlines of the new epistemology proposed later by Kuhn, Einstein's model is still static. It was Kuhn who supplied the framework within which revolutions in science could be described and at least in part understood. Kuhn distinguished the following developmental stages of science: (a) pre-science stage, what Feyerabend called theoretical pluralism; (b) normal science, where one paradigm-theory becomes dominant; (c) crisis, when difficulties are encountered and, if recalcitrant to explanation after repeated attempts, are called anomalies: (d) theoretical pluralism, where the resultant crisis ultimately leads to an atmosphere in which alternative theories are encouraged to attempt to explain the anomaly; (e) normal science, where one of these proposed embryonic theories ultimately gains acceptance and in time becomes entrenched as the next dominant paradigm-theory.[34]

Marx Wartofsky, in a recent article on creactivity, says of Kuhn:

> He failed to explain why such changes took place; and like the older philosophers of science, Kuhn also left the nature of creativity in scientific thought untouched though he fully took into account its subsequent role in science. Thus, the logical empiricists, the Popperians, and Kuhn left the question of scientific discovery to one side. They assumed it, but none of them attempted to analyze it nor to explain it. In fact, like Polanyi, Kuhn left the process of scientific change to the domain of the non-rational, if not indeed the irrational. And at least in his earlier versions, Kuhn could speak of 'paradigm-shifts' as 'leaps of faith,' borrowing Kierkegaard's irrational concept here from the philosophy of religion. Such an account, therefore, does not deal with how discovery or innovation in scientific thought comes into being, but only with the process of how such discoveries come to be accepted, how they come to be believed after they have been introduced.[35]

Even if we grant for the sake of the argument Wartofsky's claim that the emphasis of Kuhn and Popper was mostly on how new theories come to be accepted rather than on how they were arrived at in the first place, still Wartofsky's characterization glides over certain useful things that Kuhn says about the act of creation. Kuhn states, for example, that a new hypothesis or theory comes into being as a result of perceiving an old situation in a new way (e.g. perceiving a swinging stone as a pendulum or combustion as a case of something being taken from the air rather than something being added to the air). That such novel perceptions can ultimately give rise to new theories, such as Lavoisier's chemistry, is undeniable. Some have construed this initial flash of genius as perceptual; others, as judgmental. But it could be both. Perceptions involve judgments as to the "whatness" of the "thatness" in the perceptual field of vision. That it is a case of judgment is indicated by the fact that it could be wrong, a misperception. Whether it is a useful or a creative perception can be ascertained only during periods of normal science when such hunches are pursued and developed further or theoretically debugged. If all seeing is interpreting and if there is no neutral observation-language as Einstein, Hanson, and Kuhn have argued, then it would seem to follow that judgments must enter in the very beginning of the creative act.

I find also implicit in Kuhn's model of scientific development, the admission of an interval of pursuit. There are two good reasons which constitute grounds for pursuit: (1) the new paradigm must resolve the anomaly that could not be resolved under the old paradigm; (2) the new paradigm must preserve most of the problem-solving capabilities of the earlier paradigm.[36] Furthermore, Kuhn's model allows us to see better than the Einstein model that the catalyst to scientific innovation is the anomaly – the incongruous observation. In a sense the incongruities between theories and experience ultimately give rise to anomalies. It seems to me that Einstein's model does not stress enough what aspect of experience gives rise to creative leaps. Kuhn does say that anomalies provide scientists with the irritant which gives rise to the jewel. Kuhn does not say, however, how the scientist must proceed to dispel the anomaly. In all fairness it should be said that neither does Marx Wartofsky. Wartofsky does say that reasoning or judgments are involved in creativity and that one must look at actual scientific practice if one is ever to understand the creative process. I agree with Wartofsky's position though he is not the first to have said it. Herbert Fox earlier attempted to go further on this point. He states in an article on creativity in science:

> To me, the creative process so called is the thinking process directed generally toward a specific goal, the solution of a problem and culminates in a synthesis of ideas which solve or advance the solution of the problem. For as long as the mechanism and elements of thinking escape our analysis and understanding so long shall we remain in ignorance of the creative process . . . the creative process is neither more nor less than productive

> thinking . . .. Thinking need not be productive, the creative process must be.[37]

Fox rightly quotes Polya, who was one of the first to have made the point that creativity or discovery involves "plausible reasoning." Polya says "anything new that we learn about the world involves plausible reasoning."[38]

The problem with all of the above suggestions is that judgments or plausible reasoning in order to lead to some concrete conclusion must utilize concepts, but most revolutionary discoveries presuppose that the concept is invented. Ofter one cannot adequately characterize some novel phenomena unless one invents a new category. Subsumption under the category constitutes an explanation and involes judgments. It behooves us to look more closely at the major types of discoveries in science in order to explore the interconnections between discoveries and creativity in science and literature.

5. Types of discoveries and their relation to creativity

One might distinguish three basic kinds[39] of discoveries: (D_1) Accidental unplanned discoveries can arise as a result of the explanation of anomalies, as in the case of ×-rays, oxygen, and the positron. The development of novel categories and theories which can remove the anomalous observation completes the process of the first kind of discovery. Lavoisier's textbook *Chemistry* affords a good illustration of such a theoretical invention, for it explained the observed gain in the weight of sulpher. (D_2) In the second kind of discovery the thought comes first, whether we call it a theory, a law, or a model. Such a "thought" requires at times herculean efforts to produce, but it is the confirmation of novel predictions that is the focus. When a novel prediction is confirmed by observation a major discovery usually results and the theory receives the strongest empirical confirmation. When confirming observations are made of phenomena that were known prior to the creation of the theory, the theory is also supported but to a lesser degree. An excellent illustration of the second kind is providedly the discovery, in 1919, that light is bent when it passes near large bodies, to the degree that Einstein's general theory of relativity predicted. In the first kind of discovery belief or commitment plays merely a backdrop role against which something appears anomalous. In the second kind belief in the model or theory is crucial, for if the commitment is missing there will be little reason for pursuing the testing of the novel predictions.

The discovery of the positron could have been an example of this second type of discovery were it not for the fact that scientists did not believe the seemingly wild computations and theoretical postulates of Dirac, such as negative energy, negative mass, and infinite charge. When enough belief or commitment is present for pursuit, new discoveries are often the result. My second example comes from chemistry. On the basis of a law and his table of elements Mendelyeev was able to predict accurately in 1869 the existence of such elements as gallium, scandium, and germanium. When Mendelyeev proposed his revolutionary system of classification it appeared strange and counter-intuitive to many chemists. Some suggested sarcastically that Mendelyeev give up his system and classify elements alphabetically. But he did not abandon his system. Here is an excellent illustration of how considerations of symmetry can lead to fruitful discoveries. Both theoretical and experimental creativity is required in this second type of discovery.

The requirement of both theoretical and experimental creativity is particularly well illustrated in the search for tachyons, the super-luminal particles. In 1962 three physicists, Bilaniuk,[40] Deshpande, and Sudarshan finally succeeded in publishing a paper in which they showed that the existence of super-luminal particles with imaginery mass does not contradict any of the postulates of Einstein's theory. There was resistance to such a particle because Einstein said that nothing travels faster than light. Great experimental ingenuity and creativity have gone into designing experiments to slow down and observe such particles at such laboratories as the Nobel Institute (Stockholm), Princeton University, Columbia University, Brookhaven National Laboratory, and others. The search for such particles will stop only if a law making their existence impossible is discovered or if the scientists run out of money. As Bilaniuk put it in a recent article, "absences without excuses are not tolerated in contemporary physics."[41]

(D_3) Laws or theories which unify apparently different phenomena, domains, classes, or fields constitute the third kind of discovery. When Galileo and Newton discovered laws of motion which for the first time unified terrestrial and celestial physics they found one of the greatest unifying laws in the history of mankind. For over a thousand years Aristotle's system had prevailed, with two systems of physics: celestial physics studies circular motions and terrestrial physics studied natural or up and down motions and violent or projectile motions.

It is possible that the third type of discovery is reducible to the second type. Both types of discovery involve the creation of theories or laws which are new, which unify apparently disparate phenomena. It is not certain, however, whether both types generate novel predictions, the defining mark of D_2.

Many observers of creativity have noted that creativity is a matter of degree. Some low level of creativity is probably a daily phenomenon with some people, maybe even most people. Hearing someone make an apropos comment in response to a situation is not unusual. Some people more so than others have a knack for saying apt things or for making apt puns. They naturally are more creative than those who rarely make appropriate observations.

Such instances of creativity are dwarfed by the creativity required in the making of any of the three types of discovery. My hunch is that the first type of discovery probably requires the least measure of creativity because there is an accidental aspect to that type of discovery. Reasoning, memory, and a good data base are necessary, but ultimately what triggers the reasoning is the unexpected, the unexplained event. Think of the case of the discovery of x-rays. The continual exposure of photographic plates in Roentgen's laboratory led him to search for the cause. It was a matter of eliminating some possible sources and finally zeroing in on the cathode-ray tube in his laboratory.[42] The reasoning required was certainly not as complex as the reasoning that went into the development of Einstein's general theory of relativity.

We have so far determined that Einstein incorporates creativity as an integral part of his philosophy. We have furthermore discussed briefly the major types of discovery and have aligned discovery closely with creativity primarily because both aim at novelty under specified constraints.

6. Heuristics for scientific discoveries

Two kinds of questions come to mind concerning discovery and Einstein's epistemological model: (1) Is it possible to go beyond Einstein's suggestion that scientific theories or laws are arrived at by imaginative leaps? Or, to put the question another way, can there be a method or even a heuristic which could be used to speed up the rather mysterious and enigmatic process of discovery? (2) Can such discoveries or creative processes in science be explicated on a rational basis? Are they arrived at as a result of a reasoning process? Similar questions can be formulated about the process of creation in literature, as I shall discuss below.

a. *Bisociative thinking*

It has become commonplace to observe that creativity in both science and literature involves bisociative or Janusian thinking. Koestler defines bisociative thinking as follows:

> I have coined the term "bisociation" in order to make a distinction between the routine skills of thinking on a single "plane," as it were and the creative act, which as I shall try to show always operates on more than one plane! The former may be called single-minded, the latter a double-minded transitory state of unstable equilibrium where the balance of both emotion and thought is disturbed.[43]

Koestler later refers to bisociation as "the sudden interlocking of two previously unrelated skills, or matrices, of thought."[44] That such interlocking often occurs in discoveries in science as well as in the creation of metaphors in literature cannot be denied. The unification of celestial and terrestrial physics by Galileo and Newton serves as the paradigm illustration of such a unification. Our question, however, is, does bisociative thinking, if we can call it thinking, answer either of the following two questions; i.e., can it serve as a method and/or does it explicate creativity as a rational process? I think that it answers neither question adequately. It is not a very fruitful method or heuristic for making discoveries in science, for one cannot randomly go around juxtaposing different categories, fields, or matrices of thought with the hope of making a significant theoretical breakthrough. The process would be very time-consuming.[45] But even more importantly, how would one decide whether something significant comes out as a result of such a juxtaposition? Above all what Koestler does not seem to notice is the fact that, particularly in science, novel original hypotheses or theories involve novel ideas or interesting proposals for interrelationships among various parameters, as in the case of $F = ma$. Juxtaposition without unifying ideas produces unrelated diversity, not a new unity.

We can already see why bisociative or Janusian thinking won't help make the creative process in discoveries understandable either. To see that two apparently unconnected fields or categories have been unified by subsumption under a more general theory or concept, as in the case of Maxwell's equations, is not to explain that the theory or concept has been arrived at as a result of a reasonable set of steps. For help along these lines the historian of science has to enter the biographical history of the scientist during the gestation period. Detailed historical reconstructions of the problem field as well as a study of the ideational exposures that the scientist underwent will help, but neither may be sufficient as an explanation of the creative process in the particular case.

b. *The role of models and analogies as heuristics in scientific creativity*

A more fruitful approach comes from the use of analogies.[46] There are two types of analogies that can serve as a heuristic for making discoveries in science. Formal analogies can be used to provide ready solutions from one field to another.[47] Thus the laws of hydrodynamics which govern the flow of water through a pipe have supplied scientists with a useful analogy for the formation of laws governing the flow of traffic.

Kenneth Arrow[48] provides another illustration of how the formal analogies to the law of gravitation supply scientists with an empirical hypothesis as to the flow of traffic and information between two cities. Scientists have been recently testing the hypothesis that the interaction between two cities is inversely proportional to the distance between them.

Another heuristic for hypothesis-formation comes from the use of neutral analogy. The Bohr model of the atom can serve as an illustration. In the solar system, here called the parent model (M_2), the spin of a planet on its axis constitued a neutral analogy for the Bohr model (M_1), because originally the question whether electrons have spin did not arise. Revolutions of planets around a fixed center provide an example of a positive analogy because electrons are also said to be revolving around a fixed center, here called the nucleus. The tremendous differences between the size of the planets and of the electrons, and the fact that planets reflect light and electrons do not provide instances of the negative analogies between the two models. Because of the many positive analogies or similarities it was reasonable to think that the neutral analogies could be converted to positive analogies.[49] If only some scientist would have explored this neutral analogy between M_2 and M_1 and formulated the electron spin hypothesis, the Zeeman effect could possibly have been explained fifteen years earlier.

Enough has been said to establish a beachhead for philosophy in the area of creativity in science. We have seen how models and analogical reasoning can be heuristic in scientific discovery. Other methods worthy of mention are the methods of modeling, computer simulation, curve fitting, and enumerative induction. The more difficult task of establishing philosophical issues in literary creativity still remains. Discussions centering on the issue of criteria of a good literary work have long been a well-recognized province of philosophy, i.e., aesthetics. Questions of creativity, however, have often been relegated to a branch of psychology.[50] The task before us then is to explore philosophical issues in literary creativity and to show that there are significant parallels between scientific and literary creativity and achievements.

7. Metaphors – Their nature and their role

The bridge between creativity in science and literature lies in the use of models, analogies, and metaphors. Part of the reason why the sciences and humanities have for so long been seen as separated by an unbridgeable gulf is that models, analogies, and metaphors have not received adequate attention in the past. In a literature search I conducted for the preparation of this paper, the Social Citation Index (SSCI) data base confirmed a lack of attention to the cross between metaphor or analogy and creativity. There were 294 entries under metaphor or analogy and 2115 entries under creativity or creative process or creative product or innovation or discovery. But when the first and second categories were crossed there were only four entries, of which only two were relevant to my topic.

That good metaphors are among the central achievements of poetry and other genres of literature needs no argument. In order, however, to gain an insight into the process of creating good metaphors and their relation to science we should pause briefly to consider the controversial topic of the nature of metaphor.

a. *A brief historical excursion*

It was Kant and later Nietzsche who glorified metaphor and recognized it for what it is. Kant recognized that metaphors enable man to engage in thought when there are no ready literal terms or concepts available for man's use.[51] Nietzsche went even further. He saw metaphors as pervasive in all human thought and as crucial to our knowledge of the world. Metaphor is a process in terms of which man confronts the world: "A nerve stimulus, first transformed in a percept! First metaphor! The percept again copied into a sound! Second metaphor!"[52] Nietzsche becomes even bolder when he claims that originally all experience of reality is metaphorical. He states:

> What therefore is truth? A mobile army of metaphors, metonymies, anthropomorphisms: in short a sum of human relations which become poetically and rhetorically intensified, metamorphosed, adorned and after long usage seem to a nation fixed, canonic and binding; truths are illusions of which one has forgotten that they are illusions; worn-out metaphors which have become powerless to affect the senses . . ..[53]

These very suggestive ideas of Nietzsche were mostly ignored in Anglo-American philosophy. But there is a new revival of the study of metaphor as well as of Nietzsche's philosophy.

In the twentieth century much that has been claimed by poets, writers, and even philosophers about the nature of poetry and the concept metaphor has often turned out to be more puzzling than illuminating. Take for example the statement by the philosopher-poet, Wallace Stevens. He wrote that "reality is the central reference for poetry" and thus in order to understand poetry one has to "examine one of the significant components of the structure of reality – that is to say, the resemblance between things."[54] The belief that man has access to reality independent of language and of theoretical frameworks was a cronerstone of the thinking not only of some poets like Wallace Stevens but also of many philosophers of the forties and fifties. In the sixties and seventies philosophers and historians of science questioned that assumption. As a result, Stevens' claim appears naive to us today. Stevens' claim would be more plausible were we to interpret him as speaking of experience, of reality as experienced. Then the poet rightly speaks of reality structured, that is, of reality as interpreted and filtered by his concepts, desires, wishes, etc.

When Stevens speaks of metaphor he does say things that are much more illuminating. He breaks away from the classical Aristotelian conception of metaphor as based on previously existing resemblances or similarities[55] and expands the idea of metaphors to include created resemblances. Stevens states:

> In metaphor (and this word is used as a symbol for the single aspect of poetry with which we are now concerned – that is to say, the *creation of resemblance by the imagination*, even though metamorphosis might be a better word) – in metaphor, the resemblance may be, first, between two or more parts of reality; second, between something real and something imagined or, what is the same thing, between something imagined and something real as, for example, between music and whatever may be evoked by it; and, third, between two imagined things as when we say that God is good, since the statement involves a resemblance between two concepts, a concept of God and a concept of goodness.[56]

But Stevens still depends heavily on the concept of metaphor as based on resemblances even if they created rather than found. That resemblances are the basis of many metaphors is obvious. Consider, for example, the poem of H.D. Doolittle where the resemblances between waves and pines are highlighted:

> Whirl up, sea –
> Whirl your pointed pines,
> Splash your great pines
> On our rocks.[57]

What has gone unnoticed until very recently is that there are metaphors which are not based on similarities.

b. *Metaphors as experiential gestalts*

A more fruitful approach to metaphor has been developed recently, under which a new metaphor is best understood as a type of perspectival seeing. The metaphor enables us to see old objects, experiences, in a new way. Psychologists describe similar experiences as gestalt switches of the duck-rabbit type. This suggestion was developed originally by Hester,[58] following a suggestion of Wittgenstein, and more recently by Lakoff and Johnson.[59] They claim that actions, events, and objects are best understood in terms of "experiential gestalts," structured meaningful wholes within experience. Each gestalt consists of various recurring subpatterns of the whole structure.

In the case of "war" as in many other gestalts there are such aspects as participants, adversaries, stages, causation, and purpose. Johnson gives a very detailed analysis of the metaphor "argument is war" in terms of such gestalts. He shows how the two gestalts of argument and war are fused into one. After such a fusion we can use the language of war to speak of other aspects of arguments: "He attacked the weak points in my argument," or "I defended my position and destroyed his." In commenting on such examples Johnson summarizes the central aspect of this analysis of metaphors: "The metaphorical meaning is based upon the projection of one common gestalt structure (e.g., war) onto another (e.g., arguments). What emerges is a new gestalt that restructures aspects of our experience, thought, and language."[60]

What is fascinating for philosophical purposes is that metaphors on this analysis could be processed rationally. That is, one could show by what series of steps one goes from the old category to the new category created by the metaphor. Whether this interpretation of metaphor proves capable of analyzing all types of metaphors remains to be seen. What is interesting for our purposes is the usefulness of such an analysis of metaphor for the analysis of poetic creativity and for seeing the close parallel between poetic and scientific creativity.[61]

c. *Metaphors and creativity in literature and science*

Both scientists and poets often describe achievements in terms of accidental discoveries. C.S. Lewis, in describing the creation of metaphors, states: "One of the ways we invent a new metaphor is by finding it, as unexpectedly as we might find it in the pages of a book, and whatever is true of the new metaphors that we find in books will also be true of those that we reach by a kind of lucky chance, or inspiration."[62]

What the poet or the scientist finds by luck or accident is a new aspect of things, of events and experiences. These novel aspects previously went unnoticed and it is the poet who by the metaphor highlights or even creates such novel aspects. True, in the case of accidental discoveries in science there was in the beginning the perceived anomaly, the strange experimental result which could not be explained within the context of the old theories. One could say that in the case of this type of accidental discovery there is the puzzle or problem (e.g., phlogiston theory predicts a loss in weight during combustion while the experimental results show a gain in weight). But in the case of the discovery of metaphors it is not clear that a problem exists. For the most part there is no analogous anomaly for the poet. There is the problem of creating a good poem or a novel, but this is not a fruitful analogue. I suspect, however, there are times when the poet has the perception of some novel aspect before he has the metaphor, and the metaphor constitutes the linguistic characterization of his perception. Since there are no ready-made terms in the language that fit the perception or experience, the poet creates the metaphor for such purposes.[63] Ina Loewenberg puts the point in the following way: the "Metaphor increases what we can significantly say that is novel with the vocabulary we have. In the sense in which an extension of knowledge always involves the assimilation of new insights to old, metaphor extends our knowledge."[64]

d. *Gestalt-switches in scientific and literary creativity*

Both the poet's and scientist's novel experience have been characterized as a new gestalt, a seeing of old objects and experiences, in a new way. Kuhn[65] has suggested that new gestalt switches can give rise to revolutionary models and theories in science. To support his claim he cited such examples as seeing the sun as stationary and the earth as moving or seeing the swinging stone as a pendulum. For the suggestion that some theories originate in a gestalt switch Kuhn was viciously attacked. But if gestalt switches occur as well in the case of good metaphors and if metaphors can be processed in an illuminating way as specifiable shifts in the gestalts associated with this or that concept, then Kuhn's suggestion need not be looked at as irrational or obscurantist. True, the gestalt switch is not the theory, but at times it can give rise to the development of a new theory.

I do not want to give the impression that the experience of a gestalt switch precedes every novel metaphor or revolutionary theory. But I hope no one will deny that examples of such switches can be found in the history of both science and literature.

The same difficulties that scientists and others have with the proper interpretation of certain types of strange phenomena can be found even more often among students of these fields. For students novelty is relative to their state of knowledge rather than to the state of knowledge of the profession. But the difficulties are the same. Take the following passage quoted by Hanson which described a common experience of novices in science:

> An object is frequently not seen *from not knowing how to see it* rather than from any defect in the organ of vision . . .. (Herschel said) "I will prepare the apparatus, and put you in such a position that (Fraunhofer's dark lines) shall be visible and yet you shall look for them and not find them: after which, while you remain in the same position. I will instruct you *how to see them*, and you shall see them, and not merely wonder you did not see them

> before, but you shall find it impossible to look at the spectrum without seeing them. (Babbage, *The Decline of Science in England*; R. Clay, London, 1830).[66]

Even though Herschel does not use the term gestalt switch, he certainly has the gestalt switch experience in mind in the above passage. What is revealing for our purposes about Herschel's description of gestalts is that directions can be given for helping to see these patterns. The gestalt experience is therefore not irrational. Newton-Smith points out that, whereas in the duck-rabbit gestalt switch one can go back and forth, alternatingly seeing the duck and rabbit, there is no back-sliding in science. "For the innovative scientist it is much more likely that the new 'set' once grasped is fixed."[67]

In spite of this disanalogy I do feel that the gestalt switch is an appropriate characterization of the initial experience of some innovative scientists. Whether the term gestalt switch is also appropriate as a characterization of theory change or even more global changes in world-view is another matter. We have to interpret gestalt switches in this larger theoretical context metaphorically. There are some dramatic differences in the beliefs of the scientist before and after a paradigm-theory change, but such change normally undergoes a long preparatory stage of weighing reasons and looking at the data.

Something like the characterization by Herschel also goes on in the case of metaphors in poetry. When a reader does not understand a metaphor he can be directed to notice aspects of connotative meaning of the terms of the metaphor and often this will suffice for him to understand the metaphor. Much depends on the background knowledge and experience of the reader. At times one needs detailed biographical data about the poet's experience to help unravel a very personal metaphor. Such personal data are at times necessary to unravel some of Wallace Stevens' poetry.

8. A heuristic for literary creativity

a. *Role of models in literary criticism*

Thus far I have talked about literary works and the metaphors contained therein as if they could be understood in isolation, without reference to the rest of the writer's corpus. In spite of the fact that some poems and metaphors are amenable to that kind of atomistic interpretation, the formalists and the phenomenlogists have been unsuccessful in providing the full significance of some metaphors as well as of some larger literary works.

A very fruitful approach to the study of literature has come from the structuralists,[68] who subscribe to the principle that nothing has meaning in itself. When this principle is applied to literature one gets the view that in order to understand a poem or novel more deeply one has to uncover the model of the world implicit in the corpus of writings of the particular writer. Poems and novels can be read on many levels. Often the deeper significance of the work will be hidden unless one is able to reconstruct the model of the world implicit in the works. This will be particularly true of such works as Swift's *Gulliver's Travels*, Lewis Carroll's *Alice in Wonderland*, Kafka's *The Trial*, and Zinoviev's *The Yawning Heights*. Each of these works may be read on various levels. But unless we reconstruct the model implicit in the work together with the system of codes we will not decipher the deeper meaning of these works. George G. Grabowicz,[69] of Harvard, has recently provided an excellent example of such a deeper reading of the Ukrainian bard Shevchenko.

The case of Shevchenko is interesting because for almost one hundred years his poetry has been read by the masses on a very superficial level as handy bits of sentiment. Grabowicz uncovers the model that supplies the code for interpreting Shevchenko's poetry on a deeper level. Without Shevchenko's model of Ukraine in the background many of his poems would be read as they have been read on the literal level. The Zaporozhian Kozaks would not be symbols of death but simply the brave warriors that defended Ukraine against such enemies as the Turks. Tartars, Poles, and Russians. Just as in science models serve the purposes of interpretation of theoretical terms, so in literature we see that models have a parallel function. They supply the reader with an interpretation of the symbolic nature of various objects referred to in a literary work.

b. *Important similarities between the use of models in literature and science*

An interesting question raised by the structuralists' discussion of myth and models in literature is whether such modes of thinking are similar to those utilized in science or are radically different. Some structuralist literary critics like Grabowicz argue that "Mythical thought is the opposite of rational, analytical, historical thought, which takes a discrete body of data, that is, events, and by analysis and deduction sees a pattern or meaning or structure in them."[70]

In contrast, mythical thought "moves from structures to events: one starts with a structure, which in the case of Shevchenko is a sense, an understanding, a deeply felt 'truth' of, say, the nature of Ukrainian existence, and from this one creates or adapts various events or figures, for example an archetypal Cossack, or a purported historical event."[71]

This contrast between mythical and literary thinking on the one hand and scientific, rational thinking on the other has been seriously challenged in the last twenty years. The challenge has taken two forms. Philosophers like Paul

Feyerabend in his work *Against Method*[72] have argued that science proceeds counter-inductively, that "anything goes" is a better maxim for science than any of the other identifiable methodologies. Furthermore, since scientific, mythological, and literary thinking does not follow a method, it therefore contains an irrational element. The other group claims that mythological and literary thinking resemble in certain essential respects scientific thinking if only because the theories and models in the former are creations or inventions of the mind. Conversely, models and theories in science also are idealizations and contain a mythical component. They also highlight only certain features of the world to the exclusion of others. For example, in the Newtonian theory consider how even large heavenly bodies are construed for purposes of calculation as point masses, or how force is computed by abstracting from all other properties of objects except mass and acceleration. The social sciences are no different here from the natural sciences. Theories and their corresponding models highlight only certain features to the exclusion of others. Consider for example how in Marx's theory only the economic forces are taken as determinants of the superstructure of man's ideas, values, moral codes, etc., or how Freudian psychoanalytic theory takes the sexual drives as determinant of man's behavior.

None of these features appear to me to make science irrational; nor have they made it irrational for the majority of the top scientists including, as I have shown. Einstein. Furthermore, there is something counterintuitive in the claim that poetry is irrational. It is hard to reconcile such a claim with the hard cerebral reasoning that often goes into deciphering a metaphor or a poem. I therefore am prepared to defend the second interpretation of both science and literature, which claims that the two resemble each other in certain essential respects.

My quarrel then with the structuralists is not with their fruitful approach to the interpretation of literary works but with their characterization of the contrast between mythical and scientific thinking. I do not want to say that there is no difference between these two kinds of thinking but that the difference has to be put another way. Part of the difference between scientific and literary creativity lies in the different values and goals implicit during certain historical periods in each activity. Whereas predictive accuracy and control of nature are foremost values essential to any good modern scientific model or theory, certain other values dominate in good literary works within certain traditions. But this is a topic that merits separate consideration.[73]

c. *A conjecture*

More central to the question of creativity in science is to see how scientific models and analogical reasoning based on the models can serve to generate fruitful hypotheses and the further development of the theory they represent. I want now to suggest that models of the world serve as a heuristic for the literary writer or poet as well. Such models of man, of nature, of the world can provide a stimulus for the creative outpouring of the poet, novelist, or dramatist. Without such models the author is totally cut off from a perspective or an interpretation. He is in short confronted with an infinity of possibilities and it is therefore natural to wonder how one could make sense of the creative process in the face of such unlimited possibilities. The greatest writers had such models of the world to guide their creative energies, but unless the writer kept a diary or explicitly directed our attention in the work it is often difficult to get at such models. In the case of Joseph Conrad the matter is easier, for in the preface of his novel *Nostromo* he describes how he came upon the basic model of man that guided his writing of the novel. He describes how he once heard a story about an "unmitigated rascal" who had stolen a large quantity of silver. Conrad says, "I did not see anything at first in the mere story." Then: "It dawned upon me that the purloiner of the treasure need not necessarily be a confirmed rogue, that he could even be a man of character." This idea that a man or a country could be both good and evil Conrad developed in his novel. As he says, "It was only then that I had the first vision of a twilight country . . . with . . . its high, shadowy sierra and its misty campo for mute witnesses of events flowing from the passions of men short-sighted in good and evil. Such are in very truth the obscure origins of *Nostromo* – the book. From that moment, I suppose, it had to be."[74]

Models also function in other ways in poetic or literary creativity. The form of a poem or other literary work can provide a strong direction for a writer's creative energies. In such cases an element of imitation plays the central role. The poet is consciously trying to duplicate the success achieved by a particular stanza structure in another poem by another poet. Here is what Robert Lowell says about the creative process of the formation of his poem "Skunk Hour":

> The dedication is to Elizabeth Bishop, because re-reading her suggested a way of breaking through the shell of my old manner. Her rhythms, idiom, images, and stanza structure seemed to belong to a later century. "Skunk Hour" is modeled on Miss Bishop's "The Armadillo," a much better poem and one I had heard her read and had later carried around with me. Both "Skunk Hour" and "The Armadillo" use short line stanzas, start with drifting descriptions and end with a single animal."[75]

Such imitations of a work do not constitute an isolated case either in literature or in science. Kuhn has described a similar role for Newton's *Principia* or Lavoisier's *Chemistry*. Achievements such as these functioned for many years as heuristics for further research. I have elsewhere called them paragons of science on the analogy with moral paragons of virtue.[76]

In literature the writer does not bring the products of literary creativity out of nothing. Creation in literature is closer to the Platonic conception of creation found in the *Timaeus* rather than the creation *ex nihilo* found in the Bible. In the *Timaeus* Plato describes how the Demiurge creates new forms out of chaos but not out of nothing. Similarly, writers create new poems, novels, and plays out of bits and pieces they find in their environment, some story in the newspaper, part of some place they recently visited.[77] All such items are put together in a new way to create some mood, express some message, create some metaphor. Literary creations like scientific creations are further guided by a set of constraints. Among literary constraints are such things as (a) a vague something the writer wants to say, (b) semantic and syntactic conventions, (c) artistic conventions.[78]

NOTES

* Background research for this paper was supported by two National Endowment for the Humanities Summer Seminar Grants: "Model as Metaphor," University of Notre Dame, 1976 and "The Interpretation of Scientific Change," University of Maryland, 1980. The paper was written in 1982 during the period of a Research Leave granted by LaSalle University.

I am particularly grateful to Ernan McMullin for stimulating discussions on the nature of metaphors and model in the summer of 1976. I would also like to acknowledge helpful discussions on Einstein's philosophy with Dudley Shapere and Edward MacKinnon. I wish to thank the following people for making suggestions after reading earlier drafts of this paper: Olexa Bilaniuk, Daniel Durke, Lindley Darden, Albert Kipa, Mihailo Markovic, Judith Shapiro, Barbara Herrnstein Smith, Richard Strosser, and Marx Wartofsky.

This paper was presented at the Boston University Colloquium for the Philosophy of Science on December 6, 1983 and to the Consortium Working Group of Philadelphia at Bryn Mawr College on October 24, 1984.

1 Marx Wartofsky put it recently in terms of a paradox in which the more knowledge acquisition is reduced to a method the less there is room for discussions of creativity. See Marx W. Wartofsky, "Judgment, Creativity and Discovery" in *Scientific Discovery: Case Studies*, ed. Thomas Nickles (Dordrecht and Boston: D. Reidel, 1980), p. 8.

2 The most dramatic exception can be found in René Descartes' clock metaphor. For discussion of the clock metaphor see Larry Laudan, *Science and Hypothesis* (Dordrecht: D. Reidel, 1981), pp. 27–58. For the original text see René Descartes, *Oeuvres* ed. Charles Adam and Paul Tannery, 12 vols. (Paris: J. Vrin, 1956), 9:322.

3 For Aristotle's account of scientific knowledge see *Nicomachean Ethics*, 6.3: "Scientific knowledge is then, a state of capacity to demonstrate . . .. Scientific knowledge is of necessity. Therefore it is eternal." For Aristotle, scientific knowledge together with intuition (by means of which man arrives at first principles) constitutes the intellectual virtue of theoretical wisdom (*sophia*).

4 Newton states, ". . . whatever is not deduced from the phenomena is to be called an hypothesis; and hypotheses, whether metaphysical or physical, whether of occult qualities or mechanical, have no place in experimental philosophy" (General Scholium, *Principia*, 3). For a useful discussion of Newton's claim about deduction, see Errol Harris, *Hypothesis and Perception* (London: George Allen and Unwin, 1970), pp. 102–109. Harris (esp. chap. 6, "Deduction

from Phenomena'' trans. and rev. Andrew Motte [New York: Daniel Adee, 1848] pp. 506–07) proposes an interesting interpretation of Newton's principle based on the assumption that nature comes already interpreted and then proceeds to show how such ''derivations'' are possible. Mary Hesse shows what principles must be accepted if Newton's claim that the law of gravity can be derived from nature is to be plausible (*The Structure of Scientific Inference* [Berkeley: University of California Press, 1974] pp. 201–08).

In both Harris' and Hesse's discussions of Newton's claim the phenomena are not uninterpreted, but are on the contrary highly theory-laden. Naturally, if enough principles are assumed then derivation is possible. But is this really derivation from nature?

5 Carl G. Hempel, *Philosophy of Natural Science* (Englewood Cliffs, NJ: Prentice Hall, 1966), pp. 11–18. For a discussion of the problems with inductivism see also my ''Scientific Revolutions: A Philosophical Critique of the Theories of Science of Thomas Kuhn and Paul Feyerabend,'' Diss., University of Pennsylvania 1969, pp. 96–101.

6 For computer-simulated laws se Gary F. Bradshaw, Patrick W. Langley, Herbert A. Simon ''Studying Scientific Discovery by Computer Simulation,'' *Science*, 2 Dec 1983. They have developed computer programs named Bacon 1–4 that simulated data-driven inductions of such laws as Black's law (p. 973), Kepler's third law, and others.

7 See Hesse, *The Structure of Scientific Inference*. There are many features of Hesse's system which are commendable, not the least of which are: she recognized the analogical extensions of the meaning of terms from the old system to the new (''new situations to be described by complex combinations of old predicates,'' p. 221); she allows for the introduction of novel concepts in science; she recognizes a network model of theories; she stresses the importance of models in the growth of science; and she recognizes the underdetermination of theories by the data.

8 Locke's animosity towards figurative language is captured in the following passage:

> But yet if we could speak of things as they are, we must allow that all the art of rhetoric, besides order and clearness; all the artificial and figurative application of words eloquence hath invented, are for nothing else but to insinuate wrong ideas, move the passions, and thereby mislead the judgment; and so indeed are perfect cheats. *Essay Concerning Human Understanding*. 3.10.34; quoted and discussed in *Philosophical Perspectives on Metaphor*, ed. Mark Johnson (Minneapolis: University of Minnesota Press, 1981), pp. 12–13.

9 For a fresh and inciteful account of Reichenbach's distinction between the context of discovery and the context of justification, see Martin Curd, ''The Logic of Discovery: an Analysis of Three Approaches,'' in *Scientific Discovery: Case Studies*, ed. Nickles, pp. 201–21; also, a good summary of the results is provided in Thomas Nickles, ''Introductory Essay: Scientific Discovery and the Future of Philosophy of Science,'' in *Scientific Discovery, Logic, and Rationality*, ed. Thomas Nickles (Dordrecht and Boston: D. Reidel, 1980), pp. 8–10. Both Curd and Nickles have shown that Reichenbach's distinction has been misconstrued. Reichenbach intended the distinction between the context of discovery and the context of justification to be not temporal, but only logical. Thus it is compatible with Reichenbach's views to see justification as taking place throughout the process of discovery.

10 See V. I. Lenin, *Materialism and Empirio-Criticism*, 2nd. ed. New York: International Publishers, 1927). Lenin says, ''At present let us note that Mach explicitly states here that things or bodies are complexes of sensations, and that he quite clearly sets up his own philosophical point of view against the opposite theory which holds that sensations are 'symbols' of things (it would be more correct to say images or reflections of things). The latter theory is *philosophical materialism*'' (p. 33). Or again in his account of matter Lenin states, ''Matter is a philosophical category designating the objective reality which is given to man by his sensations, and which is copied, photographed and reflected by our sensations, while existing independently of them'' (p. 128). For an insightful critique of Lenin's views, see David-Hillel Ruben, *Marxism and Materialism: A Study in Marxist Theory of Knowledge* (Atlantic Highlands, NJ: Humanities Press, 1977), pp. 165–199.

11 My research on Einstein's philosophy of science was in part supported by a 1978 NEH Summer Seminar Grant. For a fuller treatment of Einstein's philosophy, see my ''Einstein's Philosophy

and Its Influence on Basic Concepts in the Natural Sciences'' (in Ukrainian), *Ukranian Engineering News* 33 (1982): 1–8; also, my article ''Einstein and the Problem of Theory Choice,'' presented at the University of Maryland, August 7, 1980.

12 For the complete correspondence see Albert Einstein, *Lettres à Maurice Solovine* (Paris: Gauthier-Villars, 1956). The letter is reprinted in *Einstein: A Centenary Volume*, ed. A.P. French (Cambridge MA: Harvard University Press, 1979), pp. 270–273. A useful discussion of the letter can be found in Gerald Holton, ''Constructing a Theory: Einstein's Model,'' *The American Scholar* 48 (1979): 309–40.

13 Einstein, ''Principles of Theoretical Physics,'' in *Ideas and Opinions*, trans. and rev. Sonja Bargmann (New York: Crown, 1954), p. 221. These essays are based on Einstein's *Mein Weltbild* (1934), together with later pieces.

14 Einstein, ''Principles of Research,'' in *Ideas and Opinions*, p. 226.

15 Einstein: *A Centenary Volume*, ed. French, p. 272.

16 Mary Hesse, *Models and Analogies in Science* (Notre Dame, Indiana: University of Notre Dame Press, 1966), particularly ''The Explanatory Function of Metaphor,'' pp. 157–177.

17 Albert Einstein, ''Authobiographical Notes,'' in *Albert Einstein: Philosopher-Scientist*, ed. Paul Arthur Schilpp (La Salle, IL: Open Court, 1970), p. 13.

18 Einstein, ''On the Method of Theoretical Physics,'' in *Essays in Science*, pp. 15–16.

19 These extensions of Einstein's model were originally proposed by Holton in ''Constructing a Theory.''

20 Arthur Fine in an otherwise interesting paper called ''Einstein's Realism'' which appeared in *Science and Reality: Recent Work in the Philosophy of Science. Essays in Honor of Ernan McMullin*, James T. Cushing, C.F. Delaney, Gary M. Gutting editors (pp. 106–133) overstates Einstein's empiricism. Fine does not seem to appreciate the limitations of empirically adequate theories and the need to supplement empirical adequacy with the internal perfections of a theory and other themata to help choose that theory which will be a realist theory. Fine says, for example, ''Indeed it would not be too far off if we summarized Einstein's views this way: 'Science aims to give us theories which are empirically adequate; and acceptance of a theory involves as belief only that it is empirically adequate'.'' (p. 125) or when he quotes Einstein on the need for choosing simplest theories Fine states ''Thus Einstein deflects questions of meaning and, just as we would expect, goes on to entheorize and to focus on empirical adequacy instead:'' (p. 123).

21 Einstein, *The World As I See It*, trans. Alan Harris (London: The Bodley Head, 1939), pp. 139–40.

22. Max Black, *Models and Metaphors* (Ithaca, NY: Cornell University Press, 1966), pp. 25–47.

23 Larry Laudan, *Progress and Its Problems* (Berkeley: University of California Press, 1977), particularly pp. 11–45.

24 Holton, ''Constructing a Theory,'' pp. 334–35.

25 Einstein, ''Autobiographical Notes,'' p. 23 (my emphasis).

26 Holton, ''Constructing a Theory,'' pp. 334–35.

27 Thomas Nickles, ''Can Scientific Constraints Be Violated Rationally?'' in *Scientific Discovery, Logic and Rationality*, ed. Nickles, pp. 285–317.

28 For a critical discussion of the above theses se my, ''Scientific Revolutions''; see also Frederick Suppé's introduction to *The Structure of Scientific Theories*, ed. Frederick Suppé, 2nd ed. (Urbana: University of Illinois Press, 1977), pp. 125–217.

29 For a persuasive discussion of the implications of scientific revolutions, see Thomas Kuhn, *The Structure of Scientific Revolutions*, 2nd ed. (Chicago: University of Chicago Press, 1970).

30 Stephen Toulmin has a useful discussion of the role of alternative geometrics to the problem of conceptual change in *Human Understanding* (Princeton: Princeton University Press, 1972), p. 51.

31 Toulmin, *Human Understanding*, pp. 41–43; see also Abraham Edel and May Edel, *Anthropology and Ethics* (Springfield, IL: Thomas, 1959).

32 For a statement of this problem, see Einstein's forceful claims discussed in this paper. More recently Mary Hesse has discussed the underdetermination of theories in her Inaugural Adress to the Philosophy of Science Association Meeting, 1978.

33 For the most vehement attack on Western Philosophy construed as a form of mirroring, see

Richard Rorty's *Philosophy and the Mirror of Nature* (Princeton: Princeton University Press, 1979).

34 Kuhn, *The Structure of Scientific Revolutions.*

35 Marx Wartofsky, "Judgment, Creativity and Discovery," p. 4.

36 Most of Kuhn's critics have missed this point and have made Kuhn out to be an irrationalist. For a discussion of this and other points relating to the rationality issue, see my "Scientific Revolutions," particularly pp. 44–70, as well as my, "A Rational Reconstruction of Kuhn's Model of Rationality in Science," *Proceedings of the 16th World Congress of Philosophy* (in press).

37 H. Herbert Fox, "A Critique on Creativity in Science," in *Essays on Creativity in the Sciences*, ed. Myron A. Coler (New York: New York University Press, 1963), p. 133.

38 Gyorgy Polya, *Mathematics and Plausible Reasoning*, 2 vols. (Princeton: Princeton University Press, 1959).

39 For a useful discussion of types of discoveries see: Richard Blackwell's *Discovery in the Physical Sciences* (Notre Dame: University of Notre Dame press, 1969), chap. 11; my "Discoveries as Indices of Progress in Science," *Proceedings of the Fifth International Congress of Logic, Methology, and Philosophy of Science* (London, Ontario: 1975), pp. 73–74.

40 For a summary and discussion of the search for tachyons, see O.M. Bilaniuk and E.C.G. Sudarshan, "Particles Beyond the Light Barrier," *Physics Today* 22, no. 5 (1969): 43–51.

41 O.M. Bilaniuk and J.R. Boccio, "Les Particules Introuvables," *La Recherche* 4 (1973): 1037–47.

42 For a discussion of the discoveries of X-Rays by Roentgen, see Kuhn, *The Structure of Scientific Revolutions*, pp. 58–59.

43 Arthur Koestler, *The Act of Creation* (New York: Dell, 2964), p. 36. The term "Janusian thinking" can be found in the writings of Albert Rothenberg: "In Janusian thinking, two or more opposites or antitheses are conceived simultaneously, either as existing side by side, or as equally operative" ("Creative Contradictions," *Psychology Today*, June 1979, p. 55).

44 Koestler, *The Act of Creation*, p. 121.

45 Alfred Burger in his lecture "Creativity in Medicinal Chemistry and Biochemistry," which was presented at the "Creativity and Science Conference" in Honolulu March 23–24 1985, describes just such a systematic trial and error testing of all known compounds for possible medicinal effects for the treatment of certain diseases. Sometimes years of testing or "juxtaposition" was productive and sometimes not.

46. See for example, Pierre Duhem's forceful statement on the role of metaphors: "the history of physics shows us that the search for analogies between two distinct categories of phenomena has perhaps been the surest and most fruitful method of all the procedures put in play in the construction of physical theories" (*The Aim and Structure of Physical Theory*, trans. Philip P. Wiener [New York: Atheneum, 1962], p. 96).

47 For many early examples of such formal analogies and their role in the growth of such fields as Darwinian Evolutionary Theory, see the fascinating article written in 1918 by M.I. Tuhan-Baranovkyi, "The Influence of Ideas of Political Economy on the Natural Sciences and Philosophy," *Annals of the Ukrainian Academy of Arts and Sciences in the U.S.* 13 (1973–77): 190–209 (English translation). For a discussion and evaluation of Tuhan-Baranovkyi's article, see my "Some Reflections on the Relationship between Philosophy and Economics," *Annals of the Ukrainian Academy of Arts and Sciences in the U.S.* 13 (1973–1977): 217–237.

Lindley Darden has proposed the hypothesis "that interfield connections are better than analogies as a source of new ideas," because (1) the use of interfield connections produces a more plausible theory if the interfield connections have been empirically confirmed . . ." and (2) "using the interfield connections in the construction of a theory provides a unification of that theory with other fields of science at the outset" ("Theory Construction in Genetics," in *Scientific Discovery: Case Studies*, ed. Nickles, pp. 167–68). For detailed historical examples of analogical reasoning in hypothesis and theory formation, see also Darden's "Reasoning in Scientific Change: Charles Darwin, Hugo DeVries, and the Discovery of Segregation," *Studies in the History and Philosophy of Science* 2 (1976): 127–169, particularly pp. 143–47.

48 Kenneth Arrow, "Mathematical Models in the Social Sciences," in *Policy Sciences*, ed. D. Lerner and H.D. Lasswell (Stanford: Stanford University Press, 1965), p. 150.

In a sweeping generalization Max Born claims, "All great discoveries in experimental physics have been due to the intuition of men who made free use of models, which were for them not products of the imagination, but representatives of real things" ("Physical Reality," *Philosophical Quarterly* 3 [1953]:140). A more cautionary note is expressed by Bernard Barber: models may also "be a source of blindness" ("Resistance by Scientists to Scientific Discovery," *Science* 134 [1961]: 596–602.

49 Mary Hesse popularized the notion of positive, negative, and neutral analogies in her now classic book *Models and Analogies in Science*. The first to use these notions, however, was J.M. Keynes in *A Treatise on Probability* (London: Macmillan, 1921). For a very thorough treatment of models and analogies and for a critical discussion of both Keynes and Hesse, see A.I. Ujomov, *Logicheskii Osnovy Metoda Modelyrovannia* [Logical Basis of the Modeling Methods] (Moscow: MYSL, 1971), particularly the discussion of Hesse (pp. 158–60).

50 After this paper was written I received in the mail a copy of Larry Briskman's brilliant article in which he argues for a philosophical treatment of the problem of creativity and why the psychological approach is inadequate. See his "Creative Product and Creative Process in Science and Art" in D. Dutton and M. Krausz, eds. *The Concept of Creativity in Science and Art* pp. 129, 155, 1981, Martinus Nijhoff Publishers, The Hague/Boston/London. In general I find his article very congenial to my approach.

51 For a useful historical introduction and critique of the various theories of metaphor, see Johnson's Introductory Essay in *Philosophical Perspectives on Metaphor*, ed. Johnson, pp. 4–20.

52 Friedrich Nietzsche, "On Truth and Falsity in their Ultramoral Sense," in *The Complete Works of Friedrich Nietzsche*, ed. Oscar Levy, and trans. Maximilian A. Mugge, 16 vols. (New York: Macmillan, 1924), II: 178.

53 Nietzsche, "On Truth and Falsity," p. 180.

54 Wallace Stevens *The Necessary Angel*: *Essays on Reality and the Imagination* (New York: Vintage Books, 1951), p. 71.

55 A good metaphor, says Aristotle, "implies an intuitive perception of the similarities in dissimilar" (*Poetics*, 1459a).

56 Wallace Stevens, "Three Academic Pieces," in *The Necessary Angel*, p. 72.

57 Hilda Doolittle (H.D.), "Oread," in *Modern American Poetry*, *Modern British Poetry*, ed. Louis Untermeyer (New York: Harcourt, Brace and Co., 1950), p. 356.

58 Marcus B. Hester, "Metaphor and Aspect Seeing," *Journal of Aesthetics and Art Criticims* 25 (1966): 205–212.

58 George Lakoff and Mark Johnson, "Conceptual Metaphor in Everyday Language," reprinted in *Pilosophical Perspectives on Metaphor*, ed. Johnson, pp. 286–329.

60 *Philosophical Perspective on Metaphor*, ed. Johnson, p. 31.

61 Hesse has argued forcefully for a cognitive function for both models and metaphors. She says, for example:

> Acceptance of the view that metaphors are meant to be intelligible implies rejection of all views that make metaphor a wholly noncognitive, subjective, emotive, or stylistic use of language. These are exactly parallel views of scientific models that have been held by many contemporary philosophers of science, namely that models are purely subjective, psychological, and adopted by individuals for private heuristic purposes. But this is wholly to misdescribe their function in science. Models, like metaphors, are intended to communicate. (*Models and Analogies in Science*, pp. 164–165).

62 C.S. Lewis "Bluspels and Flalansferes," in *The Importance of Language*, ed. Max Black (Englewood Cliffs, NJ: Prentice-Hall, 1962), pp. 37–38.

63 For a discussion of the role of novelty and freshness in all good poetry, see Daniel Burke, *Notes on Literary Structure* (Washington, D.C.: University Press of America, 1982) pp. 27–49. A detailed discussion can be found there of a refreshing characterization of a poet's wife as a "breathtakingly levelheaded girl." There is an anomalous quality to calling a wife "breathtakingly levelheaded." Anomalies like these can best be characterized by metaphors. This point has been brought out earlier by the Russian and Czech formalist theories of "foregrounding." See Jan Mukarovsky, "Standard Language and Poetic Language," in *A Praque School Reader on Esthetics, Literary Structure, and Style*, ed. and trans. Paul L. Garvin (Washington,

D.C.: Georgetown University Press, 1964), p. 19; see also Burke, *Notes on Literary Structure*, pp. 28–32.

64 Ina Loewenberg, "Truth and Consequences of Metaphors," *Philosophy and Rhetoric* 6, (1973): 44.

65 Kuhn. *The Structure of Scientific Revolutions*, p. 121. See my discussion of Kuhn's position on gestalt switches in "Scientific Revolutions," pp. 22–26.

66 Quoted in Norwood Hanson's *Patterns of Discovery* (Cambridge: Cambridge University Press, 1958), p. 184.

67 W.H. Newton-Smith, *The Rationality of Science* (Boston and London: Routledge and Kegan Pual, 1981), p. 118.

68 For an interesting survey of the structuralist literature and its application to philosophy of science, see Walter Creed, "Philosophy of Science and Theory of Literary Criticism: Some Common Problems," *PSA 1980*: *Proceedings of the 1980 Biennial Meeting of the Philosophy of Science Association*, ed. Peter Asquith and Ronald N. Giere, 2 vols. (East Lansing, MI: Philosophy of Science Association, 1980), 1: 131–140. Creed captures the essential elements of the structuralist approach when he states: "The structuralist is not concerned with the mystery or with its resolution, but with those elements of the narrative, the codes, that enable him to read the clues as clues and not simply as miscellaneous narrative details . . .. Learning to recognize and interpret these codes requires reading a number of novels and becoming aware if only tacitly of their existence and function" (p. 133).

69 George G. Grabowicz, "A Consideration of Deep Structures in Shevchenko's Works," in *Shechenko and the Critics, 1861–1980*, ed. George S.N. Luckyi (Toronto: University of Toronto Press, 1980), pp. 481–497. For a more thorough discussion, see Grabowicz's recent book, *The Poet as Mythmaker*: *A study of Symbolic Meaning in Taras* Ševčenko, Harvard Ukrainian Research Institute, 1982.

70 Grabowicz, "A Consideration of Deep Structures," pp. 491–92.

71 Grabowicz, "A Consideration of Deep Structures," p. 490.

72 Paul Feyerabend, *Against Method* (Atlantic Highlands, NJ: Humanities Press, 1975).

73 Authors who want to bring mythological and scientific thinking closer together include: Ian G. Barbour, *Myths, Models, and Paradigms* (New York: Harper and Row, 1974), particularly chap. 2, "Symbol and Myth"; E.R. MacCormac, *Metaphor and Myth in Science and Religion* (Durham, NC: Duke University Press. 1976). A particularly exciting discussion of myth in science can be found in Bernard d'Espagnat*A La Recherche du Réel*: *Le Regard d'un Physicien* (Paris: Gauthier-Villars, 1979).

74 Quoted in Rothenberg's, "Creative Contradictions," p. 59.

75 Robert Lowell's response to commentaries on his poem "Skunk Hour," in *The Contemporary Poet as Artist and Critic*: *Eight Symposia*, ed. Anthony Ostroff (Boston: Little, Brown, 1964), p. 109.

76 For a discussion of the role of paragons in the creation of paradigm-theories in science, see my "Scientific Revolutions," pp. 71–94. The chapter contains a rational reconstruction of Kuhn's *The Structure of Scientific Revolutions*, together with a definition of paradigms and paragons.

77 Cleanth Brooks and Robert Penn Warren describe this kind of borrowing further: "Though the poet must finally work out a form for each particular poem, this does not mean that he may not make use of elements of form handed from other poets – elements such as metrical patterns, symbols, and ways of relating images to theme, etc. Such elements when their use has become fixed and recognized, are called conventions" (*Understanding Poetry*, 3rd ed. [New York: Holt, Rinehart and Winston, 1960], Glossary s.v. "convention").

78 For a discussion of these and other constraints, see Burke, *Notes on Literary Structure*, p. 53.

REFERENCES

Arieti, Silvano. *Creativity*: *The Magic Synthesis*. New York: Basic Books Inc., 1976.

Arrow, Kenneth. "Mathematical Models in the Social Sciences," In *Policy Sciences*. D. Lerner and H.D. Lasswell (Eds.). Stanford: Stanford University Press, 1965.

Barber, Bernard. "Resistance by Scientists to Scientific Discovery," *Science*, CXXXIV (3479), (September 1, 1969), 596–602.

Barbour, Ian G. *Myths, Models, and Paradigms*. New York: Harper and Row Publishers, 1974.

Barron, Frank X. *Creative Person and Creative Process*. New York: Holt, Rinehart and Winston, 1969.

Barron, Frank X. and Calwin W. Taylor (eds.). *Scientific Creativity, its Recognition and Development*. New York: Wiley, 1963.

Berggren, Douglas. "The Use and Abuse of Metaphor," *Review of Metaphysics*, XVI, (December, 1962), 237–58; 450–72.

Bilaniuk, Olexa M. and John R. Boccio. "Les Particules Introuvables," *La Recherche*, IV, 1973, 1037–1047.

Bilaniuk, O.M. and E.C.G. Sundarshan. "Particles Beyond the Light Barrier," *Physics Today*, 22, No. 5 (1969), 43–51.

Black, Max. *Models and Metaphors*. Ithaca, New York: Cornell University Press, 1962.

Black, Max (ed.). *The Importance of Language*. Englewood Cliffs, New Jersey: Prentice Hall, Inc., 1962.

Briskman, Larry. "Creative Product and Creative Process in Science and Art," *Inquiry*, Vol. 23 (March, 1980), 83–106.

Briskman, Larry, "Creativity and Self-Development: A Reply to Michael Krausz," *Leonardo*, Vol. 13 (1980), 323–325.

Burke, Daniel. *Notes on Literary Structure*. Washington: University Press of America, 1981.

Burnshaw, Stanley. *The Seamless Web*. New York: George Braziller Inc., 1970.

Carrier, David. "Perspective as a Convention: On the Views of Nelson Goodman and Ernest Gombrich," *Leonardo*, Vol. 13 (1980), 283–287.

Coler, Myron (ed.) *Essays on Creativity in the Sciences*. New York: New York University Press. 1963.

Creed, Walter. "Is Einstein's Work Relevant to the Study of Literature?" In *After Einstein*: Proceedings of the Einstein Centenary Conference held at Memphis State University, March 14–16, 1979. Edited by P. Barker and C.G. Shugart. Memphis: Memphis State University Press, 1980.

Creed, Walter. "Philosophy of Science and Theory of Literary Criticism," In *PSA 1980, Proceedings of the 1983 Biennial Meeting of the Philosophy of Science Association*. East Lansing, Michigan, Philosophy of Science Association, 1980, pp. 131–140.

Culler, Jonathan. *Structuralist Poetics*: *Structuralism, Linguistics and the Study of Literature*. London: Routledge and Kegan Paul, 1975.

Cunningham, F. "Kuhn on Scientific Creativity: An Engelsian Critique," *Dialectics and Humanism*, Vol V, No. 3 (Summer 1978).

Darden, Lindley. "Discoveries and the Emergence of New Fields in Science," In *PSA 1978*, Vol. 1, ed. P.D. Asquith and I. Hacking. East Lansing, Michigan, Philosophy of Science Association, pp. 149–160.

Darden, Lindley. "Theory Construction in Genetics," In *Scientific Discovery*: *Case Studies*, ed. T. Nickles. Dordrecht, Holland: Reidel, 1980.

Darden, Lindley. "Reasoning in Scientific Change: Charles Darwin, Hugo De Vries, and the Discovery of Segregation," *Studies in the History and Philosophy of Science*, 7: (1976), 127–169.

d'Espagnat, Bernard. *A La Recherche de Réel*: *Le Regard d'uu Physicien*. Paris: Gauthier-Villars Bordas, 1979.

Dreistadt, Roy. "The Psychology of Creativity: How Einstein Discovered the Theory of Relatively," *Psychology*, Vol 11 (B), (1974), 15–25.

Duhem, Pierre. *The Aim and Structure of Physical Theory*, translated by P.P. Wiener. New York: Atheneum, 1962.

Dutton, D., and M. Krausz. *The Concept of Creativity in Science and Art*. Martinus Nijhoff, 1981.

Earman, John and Clark Glymour. "Einstein and Hilbert: Two Months in the History of General Relativity," *Archive for History of Exact Sciences*, 19: 1978.

Earman, John and Clark Glymour. "Lost in the Tensors: Einstein's Struggles with Covariance

Principles," *Studies in History and Philosophy of Science*, (1912–1916), 1978.
Edel, Arbaham and May. *Anthropology and Ethics*. Springfield, Illinois, 1959.
Einstein, Albert. *Essays in Science*. New York, The Wisdom Library, A Division of Philosophical Library. (Originally published as *Mein Weltbild*, 1934, and later as *The World as I See It*, unabridged). (Einstein's *Essays in Science* is abridged.)
Einstein, Albert and Leopold Infeld. *The Evolution of Physics*. New York: Simon and Schuster, 1938.
Einstein, Albert. *Ideas and Opinions*. New York: Crown Publishing Company, 1954.
Einstein, Albert. *Lettres à Marice Solovine*. Paris: Ganthier-Villars, 1956.
Einstein, Albert. *The World as I See It*. Covici, 1934.
Feyerabend, Paul. *Against Method*. Atlantic Highlands: Humanities Press, 1975.
Fizer, J. "Arche, Hyle and the Existential Mode of the Work of Poetic Creation," *Dialectics and Humaism*, Vol. V., No. 2 (Spring 1978).
Frank, Philipp. *Einstein: His Life and Times*. New York: A. Knopf, 1947.
French, A.D. (ed.). *Einstein – A Centenary Volume*. Cambridge, Massachusetts: Harvard University Press, 1979.
Ghiselin, Brewster. *The Creative Process*. New York: The New American Library, 1952.
Goodman, Nelson. *Ways of Worldmaking*. Indianapolis, Indiana: Hackett Publishing Company, 1978.
Grabowicz, George G. "A Consideration of the Deep Structures in Shevchenko's Works," In *Shevchenko and the Critics 1861–1980*, George S.N. Luckyj (ed.). Toronto: University of Toronto Press (1980), pp. 481–497.
Grabowicz, George G. *Poet as Mythmaker: A Study of Symbolic Meaning in Taras* Ševčenko. Harvard Ukrainian Research Institute, 1982.
Hempel, Carl G. *Philosophy of Natural Science*. New
Gruber, Howard, Glenn Terrell and Michael Wertheimer (eds.). *Contemporary Approaches to Creative Thinking*. New York: Atherton Press, 1962.
Gutting, Gary. "Science as Discovery," *Révue Internationale de Philosophie*, (1980).
Hanson, Norwood Russell. "Is There a Logic of Scientific Discovery?" In *Current Issues in the Philosophy of Science*, H. Feigl and G. Maxwell (eds.). New York: Holt, Rinehart and Winston, 1961.
Harris, Errol. *Hypothesis and Perception*. London: George Allen and Union Ltd., 1970.
Hanson, Norwood Russell. *Patterns of Discovery*. Massachusetts: Cambridge University Press, 1965.
Jersey: Prentice Hall Inc., 1966.
Hesse, Mary. *Models and Analogies in Science*. Indiana: University of Notre Dame Press, 1966.
Hesse, Mary. *The Structure of Scientific Inference*. Berkeley, California: University of California Press, 1974.
Hester, Marcus B. "Metaphor and Aspect Seeing," *Journal of Aesthetics and Art Criticism*, No. 25 (1966), 205–212.
Holton, Gerald. "Constructing a Theory: Einstein's Model," *The American Scholar*, (Summer 1979), 309–340.
Holton, Gerald. *Thematic Origins of Scientific Thought: Kepler to Einstein*. Cambridge: Harvard University Press, 1973.
Johnson, Mark (ed.) *Philosophical Perspectives on Metaphor*. Minneapolis: University of Minnesota Press, 1981.
Koestler, Arthur. *The Act of Creation*. New York: Dell Publishing Co., 1964.
Kuhn, Thomas. *Essential Tension*. Illinois: University of Chicago Press, 1977.
Kuhn, Thomas. "Historical Structure of Scientific Discovery," *Science*, CXXXVL (3518) (June 1, 1962), 760–64.
Kuhn, Thomas. "Logic of Discovery or Psychology of Research?" In *Criticism and the Growth of Knowledge*, I. Lakatos and A. Musgrave (eds.). Cambridge: Cambridge at the University Press, (1970), pp. 1–25.
Kuhn, Thomas. "The Essential Tension: Tradition and Innovation in Scientific Research," The *Third (1959) University of Utah Research Conference on the Identification of Creative Research Talent*, Calvin W. Taylor (ed.). Salt Lake City (1959), pp. 112–77.

Kuhn, Thomas. *The Copernican Revolution*. Cambridge: Harvard University Press, 1957.
Kuhn, Thomas. *The Structure of Scientific Revolutions*. Chicago: University of Chicago Press, 2nd edition, 1970.
Langford, Jerome J. *Galileo, Science and the Church*. Ann Arbor: The University of Michigan Press, 1971.
Lashchyk, Eugene. "A Framework for the Solution to the Rationality Problem," *Proceedings of the 6th International Congress of Logic Methodology and Philosophy of Science*, (submitted papers). Hanover (1979), Sec. 6, pp. 127–131.
Lashchyk, Eugene. "Discoveries as Indices of Progress in Science," *Proceedings of the Vth International Congress of Logic, Methodology and Philosophy of Science*, London, Ontario, Canada, Section V, pp. 73–74.
Lashchyk, Eugene. "Einstein's Philosophy and Its Influence on Basic Concepts in the Natural Sciences," (in Ukrainian), *Ukrainian Engineering News*, XXXIII (1982), pp. 1–8.
Lashchyk, Eugene. "The Nature of Scientific Paradigms," *Journal of Philosophy*. Vol. LXV, # 21, (November 7, 1968).
Lashchyk, Eugene. "Scientific Revolution: A Philosopic Critique of the Theories of Science of Thomas Kuhn and Paul Feyerabend," Ph.D. dissertation, (1969), University of Pennsylvania.
Lauden, Larry. *Progress and Its Problems*. Berkeley, California: University of California Press, 1977.
Lauden, Larry. *Science and Hypothesis*. Dordrecht: D. Reidel Publishing Company, 1981.
Leatherdale, W.H. *The Role of Analogy, Model and Metaphor in Science*. Amsterdam: North-Holland Publishing Company, 1974.
Lenin, V.I. *Materialism and Empirio-Criticism*. Moscow: Foreign Languages Publishing Company.
Lévi-Strauss, Claude. *The Savage Mind*. Chicago, University of Chicago Press, 1966.
Lewis, C.S. "Bluspels and Flalansferes," In *The Importance of Language*, Max Black (ed.). Englewood Cliffs, New Jersey: Prentice Hall (1962), pp. 36–51.
Loewenberg, Ina. "Creativity and Correspondence in Fiction and in Metaphors," *The Journal of Aesthetics and Art Criticism*, Vol. 36 (1978), 341–350.
Loewenberg, Ina. "Truth and Consequences and Metaphors," *Philosophy and Rhetoric*, Vol 6, # 1, (1973), pp. 30–46.
McMullin, Ernan, "Models and Analogies in Science," *New Catholic Encyclopedia*, (supplemental volume) (1974), 16.
Nickles, Thomas. "Scientific Problems and Constraints," In *PSA 1978*, pp. 134–145.
Nickless, Thomas (ed.). *Scientific Discovery: Case Studies*. Boston: D. Reidel Publishing Company, 1980.
MacCormac, E.R. *Metaphor and Myth in Science and Religion*. Durham, North Carolina: Duke University Press, 1976.
Nickles, Thomas (ed.). *Scientific Discovery, Logic, and Rationality*. Boston: D. Reidel Publishing Company, 1980.
Ostroff, Anthony (ed.). *The Contemporary Poet as Artist and Critic; Eight Symposia*. Boston: Little, Brown, 1964.
Pach, Susanne. "On Summetry in Nature and in the Visual Arts, *Leonardo*, Vol 14, No. 3 (1981), 235–236.
Rorty, Richard. *Philosophy and the Mirror of Nature*. Princeton, New Jersey: Princeton University Press, 1979.
Rothenberg, Albert. "Creative Contradictions," *Psychology Today*, (June, 1979), 55–62.
Rothenberg, Albert. *The Emerging Goddess*. Chicago, Illinois: The University of Chicago Press, 1979.
Rothenberg, Albert. "Visual Art: Homospatial Thinking in the Creative Process," *Leonardo*, Vol. 13 (1980), 17–27.
Ruben, David-Hillel. *Marxism and Materialism: A Study in Marxist Theory of Knowledge*. New Jersey: Humanities Press, 1977.
Schaffner, Kenneth F. "Einstein Versus Lorentz: Research Programmes and the Logic of Theory Evaluation," *British Journal of the Philosophy of Science*, Vol. 25, # 1 (1974).
Schilpp, Paul Arthur (ed.). *Albert Einstein: Philosopher-Scientist*. La Salle, Illinois (The Library

of Living Philosophers, Vol. VIII): Open Court, 1949.

Scholes, Robert. *Structuralism in Literature: An Introduction*. New Haven: Yale University Press, 1974.

Shapere, Dudley. "Discovery, Rationality, and Progress in Science: A Perspective in the Philosophy of Science," In *PSA 1972*. K. Schaffner and R. Cohen (eds.). Dordrecht: Reidel (1974), 407–419.

Shapere, Dudley. "The Character of Scientific Change," In *Scientific Discovery, Logic, and Rationality*. T. Nickles (ed.). Dordrecht: Reidel (1980), pp. 61–116.

Simon, Herbert A. *Models of Discovery*. Dordrecht: Reidel, 1977.

Simon, Herbert A. *The Sciences of the Artificial*. Cambridge, Mass.: The MIT Press, 1981.

Stevens, Wallace. *The Necessary Angel*. New York: Vintage Books, 1951.

Stevens, Wallace. *The Palm at the End of the Mind*. Selected Poems and a Play, Holly Stevens (ed.). New York: Vintage Books, 1971.

Summerfield, Jack and Lorlyn Thatcher. *The Creative Mind and Method*. New York: Russell and Russell Inc. 1960.

Suppé, Frederick (ed.). *The Structure of Scientific Theories*, 2nd ed. Urbana: University of Illinois Press, 1977.

Toulmin, Stephen. *Human Understanding*. Princeton: Princeton University Press, 1972.

Tuhan-Baranovśkyi, M.I. "The Influence of Ideas of Political Economy on the Natural Sciences and Philosophy," *The Annals of the Ukrainian Academy of Arts and Sciences in the U.S.S.R.* Vol XIII (1973–1977), # 35–36, pp. 190–209. (Translated from the Russian pamphlet published by the Socio-Economic Department of the Ukrainian Academy of Science, Kiev. 1925.)

Ujomov, A.I. *Logicheskie Osnovy Methoda Modelyrovannia*. Moscow: Mysl' 1971.

Victor, Turner. *Dramas, Fields and Metaphors*. Ithaca, New York: Cornell University Press, 1974.

Wheelwright, Philip. *Metaphor and Reality*. Bloomington: Indiana University Press, 1962.

Zahar, E.G. "Why did Einstein's Programme Supersede Lorentz's?" The *British Journal of the Philosophy of Science*, Vol 24, No. 2 and 2 (1973), 95–123; 223–262.

Part Four

ART AND ITS MYTHOLOGIES: A RELATIVIST VIEW

MICHAEL KRAUSZ

> When you produce the greatest effect upon the audience . . . Are you not carried out of yourself? Are you not carried out of yourself and does not your soul in ecstasy seem to be among the persons and places of which you are speaking?
>
> Plato

> In real art, theory does not precede practice but follows it.
>
> W. Kandinsky

I

This essay concerns the nature of creative products and the creative life from the creator's point of view, formulated within the constraints of a non-foundationalist and critical relativist epistemology. Central to the understanding of the creative life are our reflections about who we are and what activities we value. These reflections and motivations may be formulated as a personal mythology[1] that helps guide a creator's actions. We continually locate ourselves in such personal mythologies, which we periodically reformulate. Personal mythologies help to make sense of the activity of art making for the creator, and they help to locate the creator in relation to his practice. They are philosophies of creative life, and they motivate and orient the practitioner and the practice. Two general incongruent personal mythologies are considered here: a product-centered view and a process-centered view. The product-centered mythology holds that the principal aim of the creative enterprise is the production of certain types of things with certain valued properties, for consumption or admiration. In contrast, the process-

Margolis, J., Krausz, M. and Burian, R.M., (eds.), Rationality, Relativism and the Human Sciences.
 ISBN 90-247-3271-9.

centered mythology holds that the principal aim of the creative enterprise lies in the quality of the process itself; the things produced are merely vehicles for the process. One can add that the process aims at self-development of a certain kind.

These personal mythologies concern matters of motivation for the creator. A critic need not construe works in the same way as the creator for purposes of evaluation. Indeed, a critic is more likely to make sense of an artist's career by means of a product-centered methodology, since his access to the artist's process is usually fragmentary at best. While an artist may offer good reasons for preferring one mythology over another, the preference of one over another does not demonstrate the decisive unreasonableness of the alternative view. And this bears on the general philosophical nature of personal mythologies and the distinctive relativistic space they occupy.

The orientation is rationalist in the sense that we can make justifiable judgments about the quality of art works, though the basis for such judgments is not grounded in some ahistorical contextless logical space. Rather, such judgments can be made only within the terms of our received cultural setting and its practices. In the understanding of creative products and the creative life, I assume a non-foundationalism and a relativism of reasonableness, which I shall call a Critical Relativism.

Foundationalism, the view rejected here, holds that there are given facts of the matter of the pertinent practice, and that these are grounded in foundations, to which we have access independent of any historical or cultural practice. This "priveleged access" foundationalism may take either rationalist or empiricist forms. The rationalist foundationalist holds that we have direct intuitive knowledge of the way the world really is. The empiricist foundationalist holds that we have direct experiential knowledge of the way the world really is. In either form, foundationalism claims the possibility of knowledge about reality independent of the constraints of an inquiry. The foundationalist may hold that the history of cultural artifacts reveals some kind of approximation toward a convergent point, and that such an approximation will reveal a "right" way. But this foundationalism rapidly becomes an aesthetic imperialism, and it presupposes that art is essentially homogeneous rather than heterogeneous in its historical development.

In contrast, non-foundationalism holds that facts are taken rather than given. Yet certain recalcitrant or entrenched beliefs or commitments are necessary without which the nature of the practice as so constituted cannot be carried out. I make no metaphysical claims about the nature of the real, as it exists beyond the context of inquiry; I do not presume any privileged access for our knowledge claims. And in the realm of the arts and views of the creative life, I affirm a relativism of reasonableness of the sort urged by Joseph Margolis. On this view, different interpretations of a given domain may be formally incongruent, yet may also be reasonable or apt or appropriate to varying degrees. Here we may have good reasons for preferring one interpretation over another one, in spite of the fact that the affirmation of one interpretation as reasonable need not be sufficiently decisive to unseat its alternative as unreasonable. A relativism of reasonableness, or Critical Relativism, holds both for the understanding of the practice of art-making and for the personal mythologies that motivate it. I do not affirm a relativism of truth, which faces problems of self-referential inconsistency and does not help to unearth salient features of the domain of interest.

II

It would seem reasonable to begin an inquiry into creativity with an investigation of creative products, since we might regard certain persons and processes as creative in virtue of the prized products they issue.[2] This product-centered approach suggests a method: set out a list of uncontroversial examples of creative products and then cull out their characteristic features. Such a method presumes a univocal notion of "product" in the initial list. But our very understanding of what an eligible product in art is shifts in different historical contexts, so that no general and uncontentious list can be offered in the first place. This is a significant barrier to answering the question, What is creativity?, if by that question we mean to invite a totalist and strictly product-centered account.

Alternatively, I favor an approach that recognizes the variability of the concept of product and thereby the concept of creative product, one that is sensitive to explicating examples in local intentional settings. This alternative approach is necessary for reasons laid out by Joseph Margolis:

> . . . addressing ourselves to the art work *is not addressing ourselves to a well-demarcated object without reference to intentional considerations that illuminate at one and the same time the artist's craft and what is crafted* . . . art works have properties that cannot even be recognized independently of considering the features of an environing historical culture and the way in which gifted artists affect and modify that culture.[3]

What counts as an art product is conditionally circumscribed in a landscape of cultural entities. Sometimes what comes to be favored as an art product is part of a cultural background initially ascribing no such special status. What counts as an art product is determined within its intentional context; there is no fixed way to characterize "productness" independent of such contexts.[4]

This heterogeneity in the notion of product is well exemplified in numerous contemporary efforts. For example, on an exhibiting base in the Philadelphia Museum of Art sits a metal object with upward turning hooks, a "ready-made" of Marcel Duchamp.[5] It is entitled "The Bottle Rack." In an effort to de-emphasize precious objecthood, Duchamp offers non-traditional objects as products. He favors selection over production. Further, the American Action Painters redefined art products to include act-paintings. The residual paintings are the record of act-paintings, the totality of which comprises the painter's creative life. Harold Rosenberg reports:

> At a certain moment the canvas began to appear to one American painter after another as an arena in which to act – rather than a space, in which to reproduce, redesign, analyze or 'express' an object, actual or imagined . . . the big moment came when it was decided to paint . . . just to PAINT . . .. The painting itself is a 'moment' in the adulterated mixture of his life . . .. *The act-painting is of the same metaphysical substance as the artist's existence.* The new painting has broken down every distinction between art and life.[6]

Consider a Happening of Alan Kaprow in which are staged several 'episodes': someone pours a cup of tea; someone climbs a rope to the ceiling; a group of people rub each other's hands in cold cream, the sound produced being amplified.[7] These episodes are circumscribed as art objects. In Jean Charles Tinguely's self-destroying "Homage to New York",[8] destruction itself is circumscribed as part of the work.

Some conceptual artists have tried to push this idea still further by eliminating art objects altogether. Ursula Meyer reports:

> . . . often the artists defined the intentions of their work as part of their art . . .. Art is not in the object, but in the artist's conception of art to which the objects are subordinated.[9]
>
> The abolition of the art-object, typical for conceptual art, eliminates the concern with 'style,' 'quality,' and 'permanence,' the indispensible modalities of traditional and contemporary art . . .. "The detachment of art's energy from the craft of tedious object production has further implications," said Robert Morris. "This reclamation of process refocuses art as an energy driving to change perception". . ..[10]

These examples show the *openness* of the very concept of art products. The idea of art is, as W.B. Gallie says, "essentially contested": no clear and enduring set of conditions can be offered by way of a general characterization of it, since its intentional construal shifts from context to context and from period to period. Thus, any general theory of creativity that requires a general and univocal notion of art objects or creative products is suspect.

In an effort to retrieve a univocal notion of art objects or creative products, that is, to make such notions 'uncontested,' one might allow 'product' to include both precious objects and bundles of processes or events. One might allow the initial list to include quite different kinds of things and then proceed with the initial method. But that would ignore the special weighting of what "ought" to count as a product within the terms of the artworld ideologies from which the examples were taken. It would ignore, for example, Duchamp's essentially critical attitude toward precious object-centered orientations. Yet this is what an absolutist would require, for he assumes that there is or ought to be a convergence of artworld ideologies, at least as regards what is allowable as an eligible product. Alternatively, I suggest that the inquirer must take into account the emphases on different notions of product favored within the artworld ideologies considered, and I remain agnostic about any metaphysical convergences. Since productness may be variously defined in some artworlds, no comprehensive product-centered theory of creativity can go through.[11]

In an effort still to fill out a principled and contextually invariant notion of art product an absolutist might press along the following lines: allow previous objects *or* processes as offered, etc. *and* (to be sensitive to history!) couple this list with some scale of emphasis. That is, if we say (I): "Count as a product: paintings *or* act-paintings *or* concepts, etc.," we appear to be fair-minded in admitting *equal* claim to productness. But from what interpretive posture can we assume the rightness of the "equal" claim? Or, if we say (II): "Count as a product: primarily paintings and less so act-paintings and perhaps even less so concepts, etc." we face the same problem. From what interpretive posture can we assume the rightness of this "unequal" claim? What justifies (I) or (II), or any similar kind of grading, as an ideal? The issue turns, of course, on the justifiability of overarching ideals in such contexts.

To see how a foundationalist would treat the question of the identity of a work of music, let us consider Ernest Ansermet's views of the later Igor Stravinsky.[12] After Stravinsky came under the "nefarious" influence of the twelve-tone idiom, Ernest Ansermet was prompted to write his *Les Fondements de la Musique* and his *Ecrits sur la Musique*, the latter of which contains the essay, "The Stravinsky Case."[13] This essay attacks Stravinsky's musical sojourn into atonalism on both aesthetic and moral grounds. While the form of the attack is not new, it is interesting for its being so characteristically foundational.

Very broadly, Ansermet attacks Stravinsky on the grounds that Stravinsky's later musical idiom does not cohere with a standard which, as Ansermet conceptualizes it, remains outside any particular cultural setting.[14] Ansermet holds that tonality is the "dialectical law of musical feeling."

> . . . music, properly speaking . . . is nothing else but feelings experienced in tonal structures. (213) . . . the composer has no need to *decide* to write "tonally" in order that his writing may be tonal; it is enough for him to write only what he feels, since tonality is the dialectical law of musical feeling. (222)

So, the composer who violates this supposed dialectical law, one who *decides* to write atonally, is violating a kind of natural order.[15] And in his later work Stravinsky decided to write in the atonal idiom. Thus, on Ansermet's account, he violated something essentially human about musical expressions. So Ansermet charges Stravinsky for being no "musician-born."[16] He says,

> Only with the musician-born, the internal experience precedes the other, and with Stravinsky, the external precedes the internal. (213)

Stravinsky is ejected from the ranks of legitimate composers such as Frank Martin and Benjamin Britten. Stravinsky's decline is "in the *human* value of his works . . .." (219)

In contrast to this kind of foundationalism, I argue that while tonality may indeed be characteristic of Occidental music, it could not be of its essence, since tonality itself is a fairly late development in the West.[17] But more important, since the claim that tonality is characteristic of Occidental music is empirical, it can carry no import of necessity or universality of the sort that Ansermet's thesis suggests – unless, of course, Ansermet wishes to take the "open" concept of music and close it in a stipulative way. In such a case the resulting claim would be self-certifying and uninformative.[18]

Let us consider further foundationalist treatments. Gyorgy Doczi suggests that shells, clams, crabs, dinosaurs, frogs, horses, beetles, persons, and waves, for example, reflect the perfect mathematical proportions of the Fibonacci number series (1, 2, 3, 5, 8, 13 . . .).[19] Similarly, Erno Lendvai argues that the power of Bela Bartok's musical compositions depends upon the fact that his key works also reflect the Fibonacci number series, both in harmony and in rhythm.[20] Both of these thinkers begin their analyses with cases that neatly reflect their anticipated conclusions. For example, while Doczi calls attention to the relationships between the rings of cross-sectioned trees as reflecting his preferred proportions, he does not talk about the relationships of the tree trunks to their heights, for example. Why is the first case taken as more revealing of a deeper natural harmony? Or, when considering the "perfect" proportions in the outstretched frog, Doczi points to the relationship between the second toe and the rear foot; he does not consider the relationship between the first toe and the rear foot, etc. Why is the second toe favored? The same sort of question could be put regarding all of his myriad examples.

Correspondingly, in the case of Lendvai's tour de force account of Bartok, who indeed might have constructed his works in the way Lendvai describes, there is no indication why the Fibonacci proportions account for the power of Bartok's work, except that a theory of aesthetic human nature lurks: people most strongly respond to works of such proportions. But this thesis is never argued. And why is Bartok spoken of to start with? It would have been interesting to have seen a defense of a comparative analysis, implied by Lendvai, to the effect that, say, Ernest Bloch's major works, such as his *Sacred Service* or his first *Concerto Grosso* – which do not reflect such 'preferred' proportions – are all the weaker for their lack of such a construction. In both Doczi's and Lendvai's cases, the examples chosen reflect an assumption that they are more important, more revealing, more natural or more significant, assuming an unreflective theory of correspondence between perfect proportions and aesthetic response. The treatments of Ansermet, Doczi, and Lendvai are characteristically foundational, and they all suffer from an unsubstantiable interpretive posture of privileged access.

III

Now it becomes necessary to explicate more fully, within a non-foundationalist and critical relativist orientation, the logical space in which critical dialogue about art objects and personal mythologies operates. The general orientation favored here is both critical (agreeing to fallibilism) and rationalist (agreeing to the fruitfulness of dialogue toward consensus based on good reasons), without any absolutist construal of logic or ontology. Critical Relativism is critical and rational. The often cited contrast between relativism and rationality is a false one, since one can offer good reasons for holding that one of a set of competing interpretations or theories is better than another for specific aims or purposes within a broadly relativist orientation.[21]

The most characteristic counter-argument to relativism concerns its putative self-referential inconsistency. The target of this counter-argument is a maximal or Protagorean relativism, and it is a relativism of truth. Maurice Mandelbaum, for example, says that it is incoherent to argue that relativism is true, for if one did so, the truth of relativism would have to be of a non-relative sort.[22] And that would be contradictory. Alternatively, if one argued that relativism was only relatively true, what force could the recommendation to accept relativism have? Here one would have equivocated on "truth" over the various contexts of its application. But why perform such logical contortions? One need not equivocate on "truth" to fulfill the conditions of critical relativism. Why, in the first place, pick a maximal Protagorean relativism of truth? Protagorean relativism claims that all domains are on a par. On the other hand, there is nothing inconsistent or unreasonable about characterizing oneself as a relativist, holding, as Joseph Margolis does, that there are some domains in which at least two interpretations are formally incongruent, yet, to some degree, remain reasonable or appropriate or apt.[23] The affirmation of one does not decisively unseat its competitor as unreasonable or not apt. This does not require equivocating on "truth," and it leaves open critical discussions of evaluation in these pertinent domains. To be a critical relativist, one need not claim that all cognition is relative or that relativistically construed theories are on a par. And it is such logical indecisiveness of incongruent interpretations that is the most interesting and fruitful feature of critical relativism. In the domains of cultural or intentional entities, as in art, we find examples of critical relativism. Here, the affirmation of the favored theory or interpretation does not decisively unseat its competitors. To claim that one interpretation is reasonable or appropriate or apt does not commit one to denying that the competitors are, to some minimal degree, reasonable or appropriate or apt. Nor, on the other hand, does this observation deny that we can have good reasons for favoring one theory or interpretation *over* another.

Consider the following examples. The upbeat to the Allegro in the first movement of Beethoven's Symphony No. 1 is usually performed twice as fast as indicated by the score, assumed to be authentic. On one usual reading the 'correct' interpretation follows the principle: adhere to the consistency of the repeated figure (and the figure is repeated often within the Allegro). On the other non-usual reading the 'correct' interpretation follows the principle: adhere to the printed page. If one commits oneself to the principle of consistency of repeated figure, one has to say that Beethoven was incorrect or inconsistent or careless in his notation. Or, if one commits oneself to the principle of faithfulness to the score, one has to say that the work embodies an aesthetic inconsistency. A critical relativist could argue for the superiority of one or the other of these principles, without thereby claiming that the opposing view is not apt or reasonable. On the critical relativist's view, the claim that one reading is superior is not tantamount to saying that the opposing view is not reasonable or not apt. In this sense, the assertion of superiority is not decisive.

Consider J.N. Findlay's example of Gothic buildings as seen by architects of the Gothic Revival and by Palladian architects. J.N. Findlay says:

> Gothic buildings were differently seen by Palladian architects from the architects of the Gothic Revival. They were seen by the former in a heaven filled with idle mythological abstractions and not towering up to an ultimate mystery; they were seen in an atmosphere that loved daylight rather than dim

> gorgeousness, etc. Considering the intentional objects that the Palladian architects and the Gothic Revivalists had for their judgments, the judgments of both might have been well founded.[24]

We, as historical interpreters, can assess whether the Palladian or Gothic Revivalist architects' reaction to Gothic buildings was reasonable or apt, given their intentional contexts. We can assess how well the objects 'fit' their contexts, whether with perspicuity or strain, etc. We can judge how well the intentional object was, as Findlay puts it,

> stressed and seen in particular contexts, and . . . interpreted in a particular manner and played over by interests which may be peculiar to a period, an ideology, a region, a personality.[25]

And even if we were to decide that the Gothic Revivalist interpretation is better than the Palladian interpretation (or vice versa) according to their own intentional contexts, it does not follow that the alternative would be unreasonable or inappropriate or inapt. Here is a second case of logical indecisiveness.

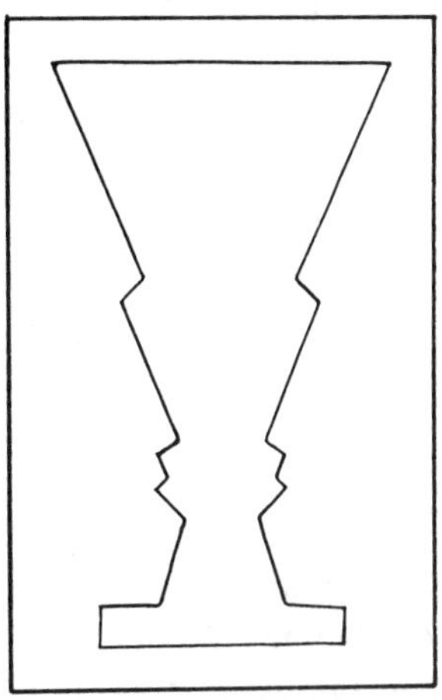

Consider yet another case, that of the face-vase figure. Figure I may be seen as a vase (the central portion being regarded as the positive space), or it may be seen as a pair of faces (the central portion being regarded as the negative space). And it may be asserted that each view may be seen with equal propriety. One view does not unseat its alternative as inappropriate. And this holds also for those versions which present themselves unequally as faces or vase. Consider Figure II.

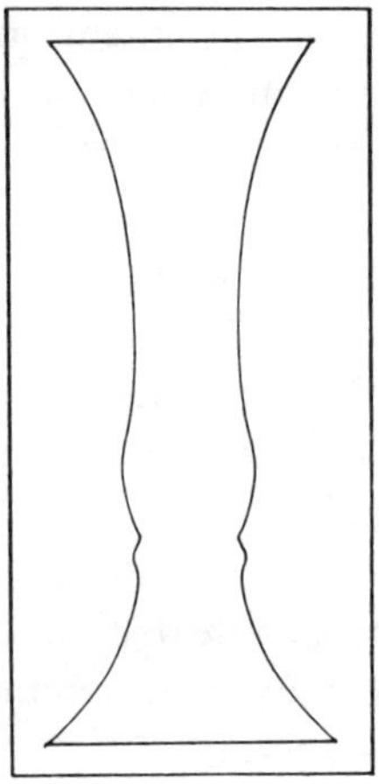

Figure II presents itself more perspicuously as an elegant elongated vase rather than as a pair of "coneheads." Yet the affirmation of the favored viewing does not render inappropriate or unreasonable or inapt its alternative viewing.

Claims of indecisiveness do not depend upon essentially unconcluded discussions. Indeed, it would be difficult to indicate what further evidence would render the discussions decisive. More strongly, in the domains considered, no further information would render the logical discussion decisive. It is not a matter of there not being enough information at hand "finally' to unseat or make inapt or unreasonable the opposing view. It might be argued, for example, that the claim of logical indecisivness in these cases rests upon an incomplete description of the surrounding conditions of interpretation.[26] A fuller historical accounting of Beethoven's intentions, aided by newly found letters, could show, it might be urged, that the principle: "be faithful to the written score" (on this occasion) does decisively show its alternative to be inappropriate, or unreasonable, or inapt. Or, a fuller accounting of the Neo-Gothic and Palladian views of Gothic buildings could show that one view decisively unseated its competitors, if, for example, newly found evidence proved that Neo-Gothic architects had no concern at all with the ultimate mystery associated with the towering constructions of the Gothic buildings. Or, a fuller accounting of the psychological contexts in which Figure II is interpreted could provide the grounds for decisively unseating one of the interpretations, by the introduction, for example, of a "conehead" aesthetic. In short, it could be objected that our examples of logical indecisiveness depend upon incompleteness of description. The issue turns on whether certain domains characteristically resist the "complete story" option. But even if the above mentioned additional stories obtained, judgments of reasonableness, etc. would still be indecisive. We could never specify how much more of such stories it would take to render the discussions decisive, for the additional stories still do not result in decisiveness. And this is a feature of the cultural or intentional domains in question.

It is important to remember that perspicuity, aptness, reasonableness, etc. are properties of intentional objects. Again, as J.N. Findlay says:

> . . . the objects in which we discover these higher-order properties are not objects as they are in nature or reality, but intentional objects, objects for us, and that means objects particularly stressed and seen in particular contexts, and in light of particular comparisons and paradigms, objects interpreted in a particular manner and played over by interests which may be peculiar to a period, an ideology, a region, a personality.[27]

At all times, the interpreter finds himself within an intentionalist context that prohibits a foundationalist self-characterization as being at the end of the story. Foundationalism presumes that an inquirer may adopt an interpretive attitude outside the historical or cultural matrix in which he interprets. It presumes an ability to formulate principles that will hold for all historical contexts, irrespective of the interpreter's cultural place.

Now does the denial that there is an accessible complete story itself presume a foundationalism? Not if one merely asserts that, whether or not there is a complete story, all attempts to formulate its method have failed; or further, that there is no need to take on the metaphysical and methodological burdens that foundationalism commits one to. Whether or not there is a complete story, one can remain silent. We needn't be atheists; agnoticism is enough.[28]

Margolis' Robust Relativism, or Critical Relativism, avoids traditional charges of self-referential inconsistency by urging a relativism that ranges not over truth but over such other values as reasonableness, appropriateness, and aptness. Yet at whatever level one asserts a critical relativism, its reasonable construal requires a logical space. Put another way, any critical relativism urged at a first-order level carries with it a second-order characterization of such reasonableness, and so on. Ever-expanding levels of logical discourse appear to approximate greater universality. But while this logical space need not be characterized as foundational or absolutist, it must be more universal than the judgment made within it. But nothing resists subsequent characterizations within an ever-enlarging historical matrix. Non-foundationalism is compatible with a 'contingent' universalism. In contrast, a universalism that claims that some features of cultures are *necessary* is unsubstantiable, since such necessity is not established by empirically discovering regularities. Even a general theory of human nature, from which such necessity might be thought to derive, will not do, since it could not be logically necessary. Universalism may be construed as an empirical claim that admits of the possibility that some conditions may, as a matter of fact, obtain in all cultures. Thus critical relativism is consistent with a ''contingent'' universalism.[29]

IV

In this section I contrast the process-centered and product-centered personal mythologies.[30] They instantiate critical relativism, and their mutual criticizability is demonstrated in their dialectical relationships developed here. The first statement takes a dialogical form, drawn from an exchange between Larry Briskman and myself, in 1974–75. This is followed by a fuller discursive explication of the process-centered view.

***

A. Creating a particular art objects is not the terminus of the process; it punctuates and helps constitute a larger ongoing process of the creative life of the artist. The point of the production of things, what motivates it, is the process, the movement, the journey. The aim of the movement is the quality of its journey, which is continuous, ongoing, and open-ended.

B. On my view, a given product is the terminating point of the process. It could then, of course, begin a subsequent creative process and then another creative product, and so on. The point of the procedure lies in the products issued. I do not see why we need to talk about any larger process.

A. One can make sense of a creative product in the terms you suggest. But my main concern is with the question, Why should an artist value making art products? If the value of the process is the objects themselves, one wants to know why one should care about them. Very few artists make enough to live from the proceeds of their work. Further, many art objects do not presume to be beautiful or especially pleasing as objects. And most aesthetic problems seem not sufficiently important to engage the lives of serious artists with the intensity that is so characteristic. The motivational question demands a deeper, more edifying response.

B. Creativity is a value notion. Calling something creative suggests that it meets some standards. In my view we can judge the quality of a work by how well it satisfies a problem situation or by how well it compares with other objects of its kind, which address themselves to similar problem situations. I don't see how you can talk of standards for your larger creative-life process. What standards would there be for that?

A. We can evaluate creative products in the way you describe. But that does not concern the motivational question. That does not speak to the artist's need to reconstitute his or her involvement with the larger process as it unfolds. The standards for the evaluation of the works, independently of the larger process, certainly pertain to the standards of the larger process. Good works rather than poor ones will affect the nature and direction of the larger life-process. But there are additonal considerations in assessing the value of the larger process. And the standards of the larger process will enter into the assessment of particular works, at least by the artist as regards the larger aims.

B. What standards are there for your larger life-process?

A. This is where we need to connect the idea of life-process with self-development. Self-development is certainly valuative, and it provides the standards in terms of which we can judge the life-process.

B. My notion of creativity is much more down to earth than yours. Mine concerns the production of creative products. Yours seems to concern the growth of the creator as well as the creative products.

A. Yes, my view concerns the growth of both the creator and his or her creative products. But to be concerned with the growth of the creator is not just an additional concern. The creator is affected by his products, and vice versa. One cannot really separate these concerns from the point of view of the creator.

B. Even so, what distinct standards would there be for evaluating your larger life-process?

A. Through self-development one becomes more alive; one is more engaged at a deeper level. One experiences things more perspicuously.

B. Are you suggesting that the more alive one is, the more self-developed one is, the better one's artistic products?

A. This is not a necessary connection, but in very general terms, the more self-developed person – at least in aesthetical pertinent ways – will produce deeper and more moving works. And it goes the other way as well. The making of good works will facilitate self-development.[31] Further, the experience of the making will help constitute the very journey of self-development. The production of things is a means of journeying. And the objects stand as markings and pointers of the journey. The art object is both the medium and the record of the self-developing experience. What motivates the movement is the developing self. The artist is part of the process.

B. But I don't see how one judges whether one has made progress in one's self-development. And I don't quite see how that would be reflected in one's work.

A. When an artist commits himself to canvas with an image in mind, what results is characteristically different from his initial intent. Then, in a backward-looking way, he sees what has been done, and then he projects the next step. In this retroactive way, he discerns the outlines of the newly travelled and emerging journey. But with that travelling, the artist gets a fuller sense of where the journey is going, and he gets a fuller sense of its emerging aim. We do not start with a clearly defined aim. It emerges as the journey proceeds. It is similar to the idea of self-development. Sometimes we sense that we have taken a wrong turn by experiencing ourselves as less functional, less alive. We experience less perspicuously. And how we feel may change in time, of course. This helps to give us a clearer picture of the interim aims along the way.

B. So there seems to be no general and abstract set of standards by which we can judge the life-process, or, for that matter, the art objects that are supposed to constitute it.

A. That is right. There are no abstract and generally applicable standards of journeys. We cannot choose between journeys from a disinterested point of view, since we are always already in some journey. There are no accessible principles that overarch between journeys, and standards are given from within the journey, including the idea of a "life process" as a value. When we say that a work is interesting or is valuable, we say so in relation to its place within a larger process.

B. But if we are ever to get anywhere, we must know what we are aiming for, even if that changes. Your view is not going to help us much when you want to know what to do next. The suggestion that the aim is in the process does not help when you are faced with the actual problem of creating something.

A. We do need specific aims within the journey. This is not to say, of course, that the journey as such is going toward a specific place, nor that such a projected place is required to motivate the journeying. The general process-centered view is not meant to provide particular aims within the journey, and for that matter, your product-centered view doesn't either. In this respect they are on a par. My process-centered view provides a motivation for the journeying.

***

The product-centered personal mythology holds that the creative process leads up to and ends with the production of an art object. And the production of a series of works can be understood in terms of a series of discrete creative processes resulting in individual products. This view does not recognize that an art object may be an integral part of a larger process that partly constitutes a creative life. A process-centered mythology emphasizes the relations between works and regards products as vehicles for the creative life. While the creative life is made so by engaging in creative acts, the products of these acts reveal something of the creative life of its artist. There is a movement to the creative life which the artist, through his or her products, documents. The creative process has to do not only with the production of things, but with the life-process from which it arises. The process-centered motivation is very different from the product-centered motivation.

Works of art take on their own life. In their making, works of art becomes autonomous of their maker. And as they emerge in the making, new, unexpected, emergent features characteristically arise. The emerging features of the work suggest a range of options for its resolution, and the resolution is as much a function of what the work demands for itself as what the maker intends.[32] Such autonomy and emergence apply to works in series as well as individual works. Often the emergent features of works in series become apparent upon viewing a one-person show or a retrospective, and the emergent features of works in series may be critical to the artist's unfolding program.

A good deal of internal dialogue goes on as to which images are to be objectified, sometimes motivated by one's personal mythology. And an art object may embody images which are not intended at all by the artist. In this regard Picasso's remarks about Braque are instructive.

> I remember one evening I arrived at Braque's studio. He was working on a large oval still life with a package of tobacco, a pipe, and all the usual paraphernalia of Cubism. I looked at it, drew back and said, "My poor friend, this is dreadful. I see a squirrel in your canvas." Braque said, "That's not possible." I said, "Yes, I know, it's a paranoiac vision, but it so happens that I *see* a squirrel." . . . Braque stepped back a few feet and looked carefully and sure enough, he too saw a squirrel . . .. Day after day Braque fought that squirrel. He changed the structure, the light, the composition, but the squirrel always came back . . .. However different the forms became, the squirrel somehow always managed to return. Finally, after eight or ten days, Braque was able to turn the trick and the canvas again became a package of tobacco, a pipe, a deck of cards . . ..[33]

Exact correspondence of an initial intended image and the image eventually embodied in the finished product is uncharacteristic, since in the objectifying, the initial image undergoes structural changes. Unexpected emergent features invariably arise. So, one cannot fully anticipate the final form of the work. This indeterminacy holds for stages toward a completed work, for completed works, as well as for works in series. Confronted with this indeterminacy, the artist often experiences anxiety in the face of the loss of his initial image, though he may often be excited by the discovery of the emerging features. Even in Representational Realism, there is no exact copy of the physically perceived situation. An artist may follow a gesture-drawing approach or a contour-drawing approach or some combination of these. And these approaches mediate as they discriminate features of the perceived world. This is in the nature of the enterprise, and is not for lack of skill. An artist could try to take up something of a mindless attitude, wishing to be overtaken by the rush of emergent features, but there can be no total randomness here. Some mediation is necessary, conscious or unconsious. The Surrealists, for example, attempted not to do away with mediation, but rather located the unconscious as the seat of mediation.[34] Even John Cage's 'random' works are not unqualified. His 'random' prints are on some sort of paper, using some sort of stylus, with some sorts of colors, employing some sort of computer program. And he shows only some of his works.

As mentioned, the process-centered personal mythology values primarily the movement rather than the resultant product, while the product-centered mythology values primarily the resultant products, which are to be evaluated independently of their origin. The product-centered theorist worries that a process-centered view results in therapy rather than art. David Carrier, for example, asks of the process-centered view, ". . . can spectators in a museum who see a painting by Pollock be said to fully appreciate the art work made by that artist?"[35] Here we must be clear about the object of evaluation. A resultant physical painting, initially produced within the context of an act-painting or a life-work context, may be evaluated in its own aesthetic terms. By suggesting that a painting may be initially conceived or motivated in a larger context, I do not rule out the aesthetic appreciation and evaluation of the physical painting on its own. The failure to distinguish a resultant object of appreciation and evaluation from its motivational context can result in the mistaken impression that the process-centered view requires that the motivational context is part of the object of evaluation.

The development of an artistic corpus and the self-development of its maker are closely connected.[36] Indeed, one may conceive of one's own life-work as self-development largely in terms of the development of one's artistic works. One may conceive of one's own becoming in terms of the becoming of one's artistic works. One may conceive of one's life-work as including the process of art making, the effect of such making on its maker, as well as the resultant product. One's life-work may involve more than the production of things. Pursuing creative activity for self-development is an exercise in defining and enlarging one's self in one's work.

Self-development is not mere change; it can be progressive. We can postdictively judge that at a prior point in our self-development we experienced more or less clearly, with more or less refinement, with more or less expansiveness, with more or less perspicuity. Self-development does not necessitate a change in the kind of art objects produced. We can experience differently and still work within a given idiom and perfect its objects. Our perfecting the objects within a given domain may reflect the changes in how we experience in general. To the extent that works reflect the quality of self-development, freezing self-development and reproducing fixed works leaves them lifeless, mechanical, and mannered.

In sum, what we make conditions how we experience, and how we experience conditions what we make. Our more perspicuous ways of experiencing give rise to more perspicuous works. It is in virtue of such perspicuity that self-development may be progressive.[37] The process-centered personal mythology, which assumes the dual aim of self-development and artistic development, may be stated as follows: artistic activity is instrumental to my self-development, and my self-development is instrumental to artistic activity. Both are valued as ends. Artistic production is motivated by self-development, and self-development is motivated artistic production.

James Munz has noted that,

> (Krausz) appears to assume that the standards appropriate for judging one's artistic production are also appropriate for judging one's self-development . . .. If artistic production and self-development are separate ends, there is no reason that their standards are identical.[38]

Munz is right to suggest that the standards for artistic production and self-development need not be identical, and where they differ, one would have to weigh one aim more heavily than the other. The process-centered personal mythology incorporating these dual aims depends upon their standards' being characteristically associated and mutually supporting.

It should be noted that setting artistic production and self-development as dual aims does not require that they should be the perpetual object of our consciousness. Artistic development or self-development could be impaired by our doing so, much as perhaps the centipede's walking might be impaired if it worried what a particular leg might be doing. Once we see, however, that there is a structural relation between artistic production and self-development, it is a matter of adjustment how much we should fruitfully dwell upon artistic production or self-development as an object of consciousness while creating.

V

No special claim has been made about personal mythology's critical function in the interpretation or evaluation of art objects. Only its place in the motivational dimension has been established. The affirmation of the process-centered mythology as reasonable – on the assumption of interest in the creative life – does not decisively unseat the product-centered view. But grounds for disagreement and mutual criticism have been provided. And, for reasons provided, foundational formulations of these personal mythologies must be held suspect.[39]

NOTES

1 This phrase was suggested to me by Roman Jakobson.
2 See Larry Briskman, "Creative Product and Creative Process in Science and Art," and Ian Jarvie, "The Rationality of Creativity," both in D. Dutton and M. Krausz, eds., *The Concept of Creativity in Science and Art* (The Hague: Nijhoff Publishers, 1981).
3 Joseph Margolis, "Emergence and Creativity," p. 7. Unpublished.
4 See Arthur Danto, *The Transfiguration of the Commonplace* (Cambridge: Harvard University Press, 1981).
5 See Ursula Meyer, *Conceptual Art* (New York: E.P. Dutton Co., 1972); and Calvin Tomkins, *The Bride and the Bachelors* (Harmondsworth: Penquin Books, 1980).
6 Harold Rosenberg, "The American Action Painters," *Art News*, 51 (1952), pp. 22–23, 48. Emphasis mine.
7 See Alan Kaprow, *Assemblage, Environments and Happenings* (New York: N.H. Abrams, 1966) as well as Alan Kaprow, *Some Recent Happenings* (Stephens Publishing Co. 1967).
8 See John Fisher, "Destruction as a Mode of Creation," *Journal of Aesthetic Education*, 8, (1974) pp. 57–65.
9 Ursula Meyer, *Conceptual Art* (New York: E.P. Dutton, 1972), pp. viii and xi.
10 Meyer, p. xv.
11 Suzi Gablik, *Progress in Art* (New York: Rizzoli, 1976). Gablik argues that the the history of art is progressive because it – or rather its non-aesthetic features – reflect or parallel the progressive features of Piagetian developmental psychology. Despite her disclaimers, Gablik is committed to making at least covert reference to aesthetic features in the history of art, for what else could qualify art history's objects? She might well have considered whether a Piagetian developmental *aesthetic* could be formulated and then see to what extent it might be mapped in the history of art. Yet such an attempt would be global and would not recognize the variability of the very idea of art or the aesthetic.
12 M. Krausz, "The Tonal and the Foundational: Ansermet on Stravinsky," *Journal of Aesthetics and Art Criticism*, 42 (1984), pp. 383–386.
13 Ernest Ansermet, *Les Fondements de la Musique* (Neuchatel, 1961); *Ecrits sur la Musique* (Neuchatel, 1971). Page references in the following text refer to Ansermet's latter volume.
14 Ansermet does qualify this by restricting his account to Occidental music. He could have used this qualification to see that in it are the seeds of a non-foundationalism. While his account remains explicitly agnostic on non-occidental musical idioms, given his theory of feelings as a variant of a theory of human nature, Ansermet is committed to pronouncing that such idioms

are not genuinely musical or not fully so.

15 It is interesting to note that Ansermet did not offer a parallel argument to the effect that mixed rhythms are unnatural by violating some dialectical law of feeling. Professor James Winn has pointed out to me that both Hindemith and Bernstein share something of Ansermet's attitude. Hindemith, who decided that tonality was the gift of nature, went back and revised his youthful dissonant pieces. And Bernstein concludes his Harvard lectures with a paean to the tonality of the earth. For a more non-foundationalist view, see Winn's discussion of Stravinsky in his *Unsuspected Eloquence* (New Haven: Yale University Press, 1981), Ch. 6.

16 The phrase "musician-born" is unfortunate since Ansermet countenances Stravinsky's earlier works. But even Ansermet's view of the early Stravinsky is not quite right, for, according to Charles Rosen, when he was writing "tonally" in the 1920's, Stravinsky was already "treating tonality as if it were an archaic and foreign language." His decision to use a severely diatonic harmonic idiom then was not aesthetically different from his decision to use serial procedures after the death of Schoenberg. Again, I am indebted to Winn for this point.

17 See, for example, Bence Szalbolcsi, *A History of Music* (New York, 1965), p. 38.

18 See Morris Weitz, "The Role of Theory in Aesthetics," *Journal of Aesthetics and Art Criticism*, XV (1956), pp. 27–35.

19 Gyorgy Doczi, *The Power of Limits* (Boulder, 1981).

20 Erno Lendvai, *Bela Bartok* (London, 1971).

21 See Michael Krausz, "Relativism and Foundationalism: Some Distinctions and Strategies," *The Monist*, 67 (1984), pp. 395–404.

22 Maurice Mandelbaum, "Subjective, Objective, and Conceptual Relativisms," reprinted in Michael Krausz and Jack W. Meiland, eds., *Relativism: Cognitive and Moral* (Notre Dame: Notre Dame University Press, 1982) pp. 34–61.

23 See Joseph Margolis, "Robust Relativism" reprinted in Joseph Margolis, *Philosophy Looks at the Arts* (Philadelphia: Temple University Press, 1978), pp. 387–401. See also "The Nature and Strategies of Relativism," *Mind*, XCII (1983); "Historicism, Universalism, and the Threat of Relativism," *The Monist*, LXVII (1984); "Relativism, History, and Objectivity in the Human Sciences," *Journal for the Theory of Social Behavior*, XIV (1984): "Objectivism and Relativism," *Proceedings of the Aristotelian Society*, LXXXV (1985), pp. 171–191.

24 J.N. Findlay, "The Perspicuous and the Poignant," *British Journal of Aesthetics*, 7 (1967), pp. 3–19. Reprinted in Harold Osborne, ed., *Aesthetics* (London: Oxford University Press) 1972, pp. 103–).

25 Findlay, Ibid., p. 103.

26 I owe this point to Professor Richard Eldridge.

27 J.N. Findlay, op.cit., p. 103.

28 It should be stressed that critical relativism requires that different interpretations addressed themselves to a common subject matter. Otherwise we would have a case of what W.H. Walsh calls 'perspectivism' instead of relativism. W.H. Walsh, *Philosophy in History* (New York: Harper Torchbooks, 1960) p. 113. According to Walsh, perspectivism holds that questions of truth or acceptability cannot be raised about perspectives since perspectives determine the descriptive terms of an inquiry and thereby the character of their domains; distinctness of perspective implies distinctness of domain. Each perspective is complete in the sense that there is no mutual translatability between them. Walsh rehearses the position as follows:

> If we are asked, 'which is truer, the Catholic or Protestant versions of the events of the Reformation?' we must reply that we cannot say. There is simply no way of comparing the two accounts, each of which is complete in itself. The Catholic looks at the Reformation from one point of view and offers his interpretation of it; the Protestant looks at it from another and produces a different interpretation. Since points of view are, ultimately, not matters of argument . . . we cannot say that one is 'objectively' better than another, and so must recognize that the Catholic and Protestand versions do not really contradict each other. . .."

The two interpretations do not contradict each other, since, on the perspectivist theory, they do not speak of the same thing. On that view, no pair of interpretations of differing perspectives can address themselves to the same domain. Now, whether or not perspectivism is defensible (and I think it is not), it is not a form of critical relativism. Critical relativism requires at least the logical incongruence of interpretations about a common subject matter. And that is denied by perspectivism.

29 For anthropological examples of the cultural relativity of perception, see Philip Kilbride, "Pictorial Depth Perception and Acculturation Among the Baganda of Uganda," *American Anthropologist*, 1969.

30 Other examples of personal mythologies, beside the process-centered and product-centered views, include Kandinsky's Theosophy or Malevich's Nihilism, or Mark Toby's Zen Buddhism.

31 Winn is suspicious about this connection, since he acknowledges that some poorly developed people (e.g. Mozart, Caravaggio, and Ezra Pound) do make aesthetically admirable objects. A quick response would be that they were creative despite such deficiencies, much as one might say that, contrary to his own view, Rilke would have written yet better poetry if he had undergone psychoanalysis after all. A more measured response would hold that the issue is essentially undecidable. Yet such undecidability of the *causal* connection should be separated from its being plausibly incorporated in a personal mythology for its motivational function. So long as the connection is not unseated, its adoption remains eligible. Indeed, Winn is right to identify this connection as one nerve which distinguishes the Romantic view from the Classical view, and certainly *that* difference instantiates the Critical Relativist view (of undecisiveness) that I have sketched.

32 This, in a nutshell, is why a pure reenactment model of historical explanation as applied to the arts is insufficient. The New Critics are right to resist the suggestion that *the* interpretation of a work of art is to be found in the intentions of the artist. But they are right for reasons not central to their argument. It is not the case that conditions 'external' to a work are irrelevant to its interpretation. Frank Cioffi has rightly argued that the very distinction of internal/external cannot be sustained. But further, the intentions of the artist characteristically do not exhaust the causal circumstances in the production of an art work. Art works are often, and most interestingly, the result of unintended emergent processes. Most basically, though, since art works exemplify critical relativism, the New Critics are wrong to search for a method which will yield *the* meaning of any work of art. The New Critics have mistakenly tried to follow the logical features of decisiveness, which they (probably wrongly) thought applied in the natural sciences.

33 Françoise Gilot and Carlton Lake, *Life with Picasso*, (New York, 1964), pp. 68–69.

34 See Andre Breton, *What is Surrealism*? (London, 1936).

35 David Carrier, Letter to the Editor, *Leonardo*, 13 (1980), p. 262.

36 Formulations of self-development, for example, are found in diverse idioms. See M. Krausz, "A Painter's View of Self-Development and Creativity," *Leonardo*, 13, (1980), pp. 143–145; Chang Chung-Yang, *Creativity and Taoism*: *A Study of Chinese Philosophy*, *Art and Poetry* (New York: Harper, 1970); Ernest G. Schachtel, *Metamorphosis*: *On the Development of Affect*, *Perception*, *Attention*, *and Memory* (New York: Basic Books, 1959).

37 In accordance with sections II and III, the very notion of perspicuity should be understood in non-foundationalist terms.

38 James Munz, Letter to the Editor, *Leonardo*, 13 (1980), p. 350.

39 Among my heuristic spirits I am pleased to acknowledge Richard Bernstein, Larry Briskman, Richard Burian, David Carrier, Noel Carroll, Constance Costigan, Robin Dintiman, William Dray, José Ferrater-Mora, Roman Jakobson, James Munz, Coral Lansbury, Hugh Lacey, Joseph Margolis, Jack Meiland, Martin Moskof, Karl Popper, Frederik Prausnitz, M.C. Richards, James Winn, and members of the Greater Philadelphia Philosophy Consortium.

ON RELATIVITY, RELATIVISM, AND SOCIAL THEORY*

BRAULIO MUÑOZ

I

Anthropological and sociological research in the last few decades has shown us a perspicuous cultural diversity undergirding a manifoldness of epistemic and value positions. These reports have added weight to the salient position within contemporary social theory which holds that relativism in all spheres of life is an obvious case of the matter. The burden of proof, it seems, now rests on the shoulders of those who would want to argue for some kind of universality regarding humans and their cultures. They must show that what seems obvious is not really so. I shall attempt here to discharge this burden by showing that indeed, the obvious relativism is not really so. To this end I draw from classical sociological theory and make a crucial distinction between *relativity* and *relativism*. That is, I acknowledge that relativity is indeed a fact of life. Not all meaning is transparent, nor are all ideological-practical positions shared by all cultures or by all individuals within a given culture – despite the strain of hegemonic structures to exact consensus. But such relativity, I hold, refers to the level of social life I call the *presented*. This level, however, does not exhaust all of social life, though it is its most conspicuous side. There is another level in social life which I call the *given*. This level has to do with more permanent and underlying social constructs both within societies as well as among societies.

II

But is not the attempt to argue for any kind of universality a serious misunderstanding of the tradition? After all, is it not the case that today's sociologists and anthropologists, heirs to classical social theory, are to be counted among the most ardent supporters of relativism? The issue warrants discussion.

During the rise of modern social theory in the middle and late 19th century,

Margolis, J., Krausz, M. and Burian, R.M., (eds.), Rationality, Relativism and the Human Sciences.
 ISBN 90-247-3271-9.

advocates of the universality of Human Nature and hence, of the equal worth of cultures, fell into two broad categories: there were those who argued for special creation, and there were those who posited a universalising thrust in Nature. The first relied on theological arguments, which ultimately approximated Plato's position regarding the relation between appearances and essences: relativity was overcome with reference to a realm outside the phenomenae world, Kant's philosophy exemplified the indebtedness of secular thought to the theological framework – the postulation of a Noumenal world being Kant's ultimate ground for universality. In the second case, Nature was seen as the ground for man's special place in the cosmos, either romantically or scientifically conceived (*generatio aequivoca*). Even Darwin's momentous work, which shattered the teleological conception of natural processes, treated humans as partaking of universal characteristics achieved through similar experiences and adaptive moves. Within religion or science, there was an explicit and decisive recourse to a fundamental conception of Human Nature. In both cases, too, human life was seen not as the ultimate datum but as an epiphenomenon.

The position of the founders of modern social theory regarding universals differed drastically from both the theological and naturalist positions. In fact, they rose to argue that both positions were inadequate for the proper understanding of human life. Marx, Durkheim, Weber, and Freud, the clearest exponents of social theory in the 19th and early 20th centuries, carried forward a tradition supported by the Enlightenment and attempted to show the Deity as a human construct. These social theorists also made us aware of the inadequacy of radical positivism, whose present-day heir is socio-biology, in accounting for social life with evident though varying degrees of freedom.[1] Herein lie the main reasons for the late 19th and early 20th centuries' understanding of social theory as an ally of relativism. Having questioned the two acknowledged grounds for universality, commentators then assumed social theory inevitably led to relativism. Today this reading of the initial thrust of social theory is a commonplace. This is particularly evident in the work of leading contemporary philosophers as well as (excepting the early Parsons) anthropologists and sociologists.[2] What is forgotten is that Marx, Durkheim, Weber, and Freud did not waver from their self-imposed task of guiding the efforts of modern man in his search for new ways to re-integrate the social fabric – that is, the fabric shattered by the social forces that brought about the collapse of the integrative role of the church. That integrative effort had relied fundamentally on the assumption of cultural universals within and between societies.

III

The works of the founders of modern social theory (i.e., Marx, Weber, Durkheim, and Freud) have one thing in common: they all evidence a determined effort to probe beneath mere appearances. They sensed a need to understand what lay underneath. For them, description was never enough. They not only welcomed universals, they actively sought them. Their search led to an elemental but crucial distinction: in social life, there are things that are *presented* and things that are *given*.

What is *presented* is what is readily at hand, the surface of human activity. For Marx, this meant superstructure, market relations, concrete political processes. The distinction is evident in his discussion of the circulation of commodities, in the first volume of *Capital*: "the coat is a depository of value," he wrote of his prime object-lesson, "but though worn to a thread, it does not let this fact show through."[3] To understand why and how the coat is a depository of value, one has to penetrate beneath the circulation of commodities (with which political economy is content), to uncover human labor as the creator of all value. For Durkheim, the "presented" meant the *idola* of the layman as well as the descriptive efforts of social investigators who fail to penetrate the veil of appearance. "Every contract," wrote Durkheim, against the utilitarians, "supposes that behind the parties implicated in it there is society very ready to intervene in order to gain respect for the engagements which have been made."[4] We find the noncontractual element of contracts only by going beneath that which appears. For Freud, the "presented" lies in consciousness open to self-reflection or in the pre-conscious open to recollection. Thus, to Freud, the difference between the psychoanalytic session and the confessional act lies precisely in that "the sinner tells what he knows; in analysis the neurotic has to tell more."[5] To uncover what lies beneath the presented is the task of patient and physician: the doctor "must turn his own unconscious like a receptive organ toward the transmitting unconscious of the patient."[6] Weber's entire effort to develop a *Verstehen* methodology for sociology was geared to penetrate the surface of meaning.[7] In all cases, that which appears is the realm of everyday human activity, the field where interests and values collide. At this level, *relativity*, more or less contained by straining hegemonic structures, is a fact of life.

The *given* admits of a fundamental distinction: what is given to us as biological organism by our natural constitution, and what we give ourselves as social beings in and through history. The founders of modern social theory were agreed that humans are natural beings possessing certain natural and universal characteristics, such as the ability to labor, speak and develop cultures. In this sense they were not relativists. But the decisive misunderstanding for attributing a relativist position to early social theory did not concern *natural* but rather *cultural* considerations. Early critics of social theory were correct in suspecting that the claims of social theory become a serious challenge to both traditional theology and modern biology when it is argued that: (a) granting a conception of the divine is pervasive in human affairs does not as such entail that conception's being dictated by an eternally given Human Nature; or (b) granting universal biological characteristics in man does not entail the reducibility of the cultural to natural processes. Such positions are, indeed, at the base of Durkheim's view of society as an entity *sui generis*, Freud's relentless effort to rescue psychology from biology during his heroic middle period, Marx's attention to history, and Weber's sociological understanding of religion.

But did the founders of social theory hold that the variety of cultures manifest an ultimately random manifoldness of human life? My reading is that they were not relativists in this sense either. They all agreed that there are *socially produced* constructs which are found in all societies. In Marx this is clear in his position regarding labor as the underlying constant in all cultures: "so far . . . as labour is a creator of use values, is useful labour, it is a necessary condition, independent of all forms of society, for the existence of the human race . . .."[8] In Durkheim this is clear in the Kantian position he adopted concerning the fundamental distinction between sacred and profane, which he believed underlies all cultural creation, including the orientational matrix of space and time. On this score Durkheim has been blatantly misunderstood by those who claim his as their relativist predecessor. Different cultures may have different conceptions of time and space but *that* they must have such categories and *that* such categories are grounded in the distinction of sacred and profane is universal.[9] Freud's image of the mind as it changed over the years was never meant to apply only to Western man. The conscious-unconscious distinction was meant to be a universal distinction.[10] Weber's contribution to historical sociology and cross-cultural analysis is based on the possibility of cross-cultural understanding both in space and time. As he put it "one need not have been Ceasar in order to understand Ceasar."[11] Indeed, the current understanding of relativism in the social sciences as a whole moves *in opposition* to the initial thrust of social theory to ground universals in cultural processes.

IV

Why such a consistent misreading of the initial animus of social theory? What is it that so offends contemprary social theorists in these early attempts? I believe the offense rests on two considerations: (a) in the belief that by going underneath the *presented* these theorists neglected the concreteness of experience. The social universals uncovered, they feel, give the impression of influencing concrete individuals as if from afar. Under these circumstances, modern theorists believe, freedom reduces to at best a freedom for the species or society, not for its individual members. Our liberal tradition resents the oversocialized view of man. Clearly, these critics are right in protesting against the oversocialized view of man we find scattered in some of the works of the founders: the primacy of Class in Marx, Society in Durkheim, Phylogenesis in Freud. However, the critics are wrong in abandoning these earlier efforts. Given the insightful and fruitful beginnings to be found in the overall thrust of the founders' work, the task should be to push forward. I shall discuss this point presently. (b) The other consideration has to do with the penchant among some earlier sociologists for adopting a kind of social teleology. August Comte and Herbert Spencer, for example, misappropriated from (Lamarkian) biological thought the idea of unilinear evolution; more recently, Talcott Parsons came full circle to revive a teleological reading of cultural evolution that nicely sets up American Culture as the epitome of civilization.[12] It so happens, it has been argued, in view of these and other cases, that what is considered universal is nothing but the Western universalized. I shall also return to this issue, below.

V

Within American sociology the reaction stemming from the first set of considerations was developed largely by followers of the Pragmatism of the Chicago tradition, in particular the influential work of G.H. Mead. In Europe, the opposition to sociological universals was led by strands of existentialism and phenomenology, which reacted against a structuralism tracing one of its roots back to Durkheim's work and another to an orthodox Marxism that gave primacy to a "base" narrowly and mechanistically conceived. These corrective moves on both sides of the Atlantic emphasized "experience" and "Lebenswelt." I shall discuss briefly G.H. Mead's contribution here.[13] It will help me bring into focus what I believe is a necessary corrective to the long search for socially grounded universals.

The Meadian interactive theory of meaning has great suggestive power, seems intuitively correct, and is supported by empirical findings.[14] Mead's understanding of the emergence of meaning borrows rather heavily from Hegel's conception of the dialectic, particularly from Hegel's discussion of Master/Slave relations in *The Phenomenology of the Spirit*. Mead argues that meaning is established in the dialectical exchange of gestures and utterances among individuals who are part of an ongoing interactive system. Hence, meaning is never "given" but always "produced" and, I would add, "presented." This means that there never is an *a priori* or frictionless agreement as to what things or events are. Individual identities, for example, are "negotiated" in a process of claim and validation among interacting individuals. In social life, not all that is individually meaningful is socially meaningful, and vice versa.

Perhaps it will be helpful to illustrate this interactive theory of meaning with the following diagrams:

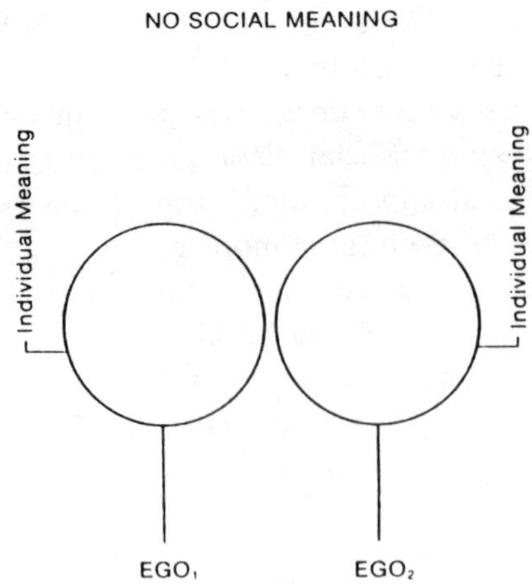

a

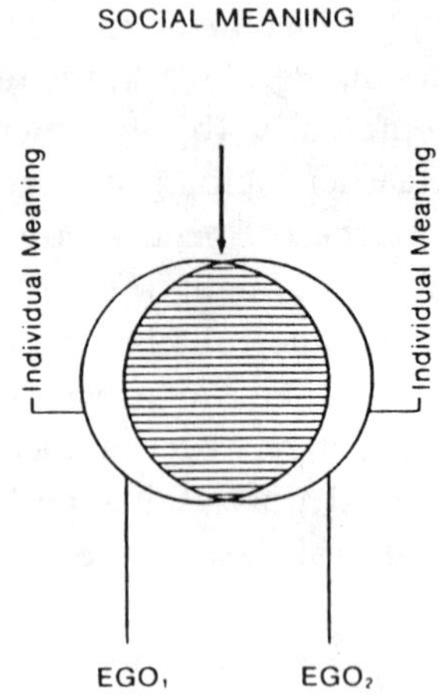

b

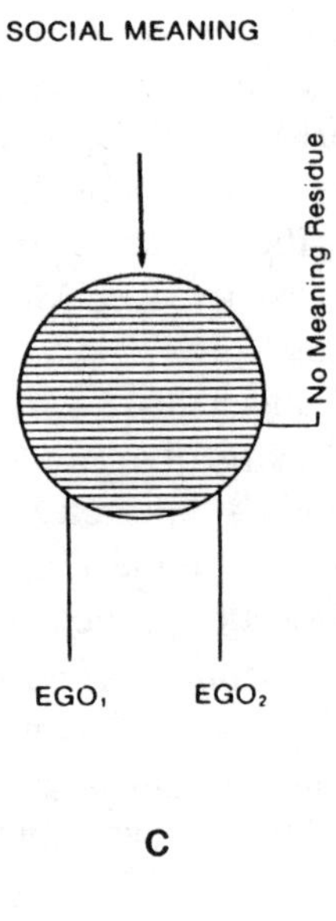

c

Situation (a) refers to total inability to communicate. This ideal-typical case approximates Feyerabend's position regarding incommensurables, Kuhn's discussion of competing paradigms and Winch's discussion of different forms of life. But, clearly, if such an extreme case existed in everyday life, neither society nor a personality could possibly exist, for these are human constructs grounded in interaction. In his recent book, Richard Bernstein has shown that in the works of Feyerabend, Kuhn and Winch, who are widely regarded as the leading voices for relativism, there is a deep acknowledgement of the impossibility of this state of affairs.[15] To take the other extreme condition, (c) points us to an impossible state of affairs as well. In this situation, what each interacting individual means would be transparent to all others. Perhaps this would approach Kant's conception of the Kingdom of Ends wherein the grounds for law, for example, would be evident at all times and imperatives would not be needed. Of course, such a state of affairs could be seen as a goal reached only asymptotically: eventually, self-reflection would entail social existence. In a fundamental way this was one of Marx's most cherished hopes. This is what he meant when he argued that communism is the beginning of "human history." It also underlies Peirce's understanding of the search for truth through reason. But in this case we are no longer situated in the extreme ideal-type.

Situation (b) is our situation as humans whenever and wherever we find ourselves. And in this case, as is clear in the diagram, some individual meaning will always be "lost." Individual personalities and societies are built only on such conditions.[16] It is the price paid for social life. There is no need to hold that interaction must always or even usually be totally transparent to the interacting parties. All that is needed for social interaction to take place is a sufficient degree of shared values, beliefs and material conditions. Furthermore, it is in the totality of such interactive occations that freedom for the individual is attained, for freedom is a social construct and, hence, to be won or lost at the level of the *presented*. Freedom is not guaranteed by either a natural or social *given*. If freedom is possible only on the condition that society does not absorb the individual, it is impossible when there are no social parameters circumscribing individual development. This position is at the core of the founders of social theory who begin with the understanding of humans as social beings through and through. Mead's retention of the Kantian transcendental "I" as the seat of free will and creativity kept him from a more sociological reading of the relationship between individual and society.

VI

There is another aspect of Meadian social thought that has been one-sidedly appropriated by modern social theorists, in the search for socially grounded universals. Contemporary heirs to the Chicago tradition claim from Mead the tenacity to remain at the level of the *presented* in interaction. Consequently, they often underplay what I would like to call the "social-construct precipitate." That is, in my view, it is crucial to maintain at the level of consciousness that the interactive process does not start spontaneously from itself, but is always grounded in structures already formed in antecedent interaction itself. Indeed, interaction is possible only on the condition of presupposed social constructs. Furthermore, the interactive process is not exhausted in its present enactment, but each interactive process leaves behind precipitates which add to or modify underlying social constructs. This, I believe, is a more faithful understanding of Mead's theme. But among some social scientists there is a desire to assert individual freedom in the face of social constructs. This leads them to overemphasize both the present and future of action, and thus to underplay the role of the past as constitutive of any activity whatsoever. It is only because action can add to or modify the underlying social constructs that support it that one can speak of "history" and "biography." These underlying social constructs are what I have called *givens* in the dual sense of the natural and the social.

VII

Both American and European corrective efforts unduly limit their analysis to symbolic, cognitive or affective elements of the interactive process. One element is often conspicuously absent: power. If meaning is negotiated, then questions must rise as to the possibility that not all individuals have an equal chance of being heard in an equal role interaction. There is no *prima facie* necessity to accept a liberal conception of social relations. It is the absence of a serious discussion of power that renders Gadamer's dialogical hermeneutics inadequate as a theory of meaning and social relations and open to attack à la Habermas. Habermas is correct in borrowing from Marx, who showed that behind the formal rights and freedoms in liberal societies, and in any society so far with similar rights and freedoms, there lie power relations among unequals. Not only the question of what gets established as socially meaningful, but that of meaning itself, is affected by power relations. This is what prompts Habermas to call for a depth hermeneutics.

The question of power leads me to the second major set of considerations that offend contemporary social theorists and contribute to their misunderstanding their own tradition. Here, too, Mead's theoretical contribution is central. It seems to me that a case can be made for the "translation" of the Meadian interactive model (with the modification noted), which applies basically to interactions within societies, to interactions between societies. In other words, in the encounter of cultures, meaning is also "negotiated." Given the historical circumstances, however, such "negotiation" has in practice meant, more often than not, ruthless imposition of wills. Nonetheless, the anthropological evidence shows that (1) no culture ever imposes itself completely on another (as in diagrams (C), and (2) no dominant culture has yet succeeded in totally fending off the influence of the dominated (as in diagram (A)). This has to do with the fact that the collision is one of claim and validation, not sheer physical clash. The master cannot but be defined by the slave and vice versa. Freud has taught us to be suspicious of vehement denials regarding the influence of the Other in our lives.

When we consider interaction among societies or cultural wholes we can readily see that the anthropological animus against nomological laws regarding cultural processes issues from a correct reading of the power relations between different cultures. The anthropologist's heart-felt demand that non-Western cultures be respected for what they are, and not judged inferior according to our standards, is in large part a moral-political position. The historical record is sufficient reason for the noble stance. But anthropologists themselves are unwilling to acknowledge this element in their plea mainly for two reasons: (a) because of their wrong-headed attempt to divorce science from value, and (b) because power relations between societies are seldom at the core of their discourse. In other words, anthropologists often lack the courage of their convictions and/or are naively liberal in their conception of cultural exchange. That Westerners have behaved in reprehensible ways should not lead us to lose sight of what is at stake. Anthropologists often retain relativism as the ultimate bastion against Western imperialism; a noble but erroneous position, for it shifts the argument from power to cognition.

The anthropologist's moral shyness and liberal naiveté play rather nicely into today's philosophical discussion of relativism. Many philosophers today, even or precisely when drawing on anthropological literature, tend to regard questions of relativism as purely theoretical-epistemological questions. This leads me to conclude that these philosophers, too, have accepted *a priori* the liberal idea of logical argumentation free from its underlying social context as the essence of the theoretical. But power more often than not refuses to submit to logical argumentation. The issues of relativity are never purely theoretical; theory, as Marx noted, is never divorced from practice. Individuals do not only "theorize" while establishing meaning in their everyday life, they live. Cultures do not clash theoretically but historically; "beach-heads" are not only cognitive maps but power expeditions.

VIII

The *given* serve as ground and parameter for everyday life. But precisely because they are ground and parameter, they do not determine the details of everyday existence. To lose sight of this would lead to unnecessary confusion. The attack on Kant for failing to show exactly how the Categorical Imperatives function in particular instances issues from such confusion. A similar confusion lies behind the modern social scientists' attack on social universals. If universals were to affect directly and in detail the concreteness of daily existence, they would indeed be too constraining. But, such demands and fears are unfounded: first, the *given* is not to be equated with universals obtaining within a given culture; secondly, the *given* do not intervene directly in concrete instances so as to lead to unfreedom. Universals obtaining only within a given culture, such as particular epistemologies, worldviews, and the like, are exemplifications of the *given*, but they are necessarily mediated through their embeddedness in a specific time and space: thus, Western culture and the *given* are not to be equated. The *given* are a-temporal in a certain sense – as far as humans at any given time in history are concerned.

But, however slowly, the *given* do change in history. I am not claiming that the *given* are eternal or that they obtain in all possible worlds. Whatever the nature of the *given*, it is the case that they refer only to our development so far. Humans can and do change the *given* in and through history; this position is at the core of the tradition I draw on. But such changes cannot be brought about by one political move nor by one generation. Changes in the *given* are analogous to changes in language: in language, changes in *parole* are frequent; changes in *langue* are much less so. In this sense it is quite meaningful to speak of human universals grounded in history so far. There is no need to see the question of universals as an either/or proposition: either there are eternal universals or there are no universals at all. Furthermore, it is quite vacuous to argue whether universals obtain in all possible worlds. The only world that is important to us is the one we live in with its historical sediment and its horizon towards which our hopes and actions lead. Objections to this view of universals, which follow Popper's lead regarding knowledge without a knowing subject for example, are interesting but not relevant.

IX

If relativity leads not to relativism but to a socially grounded universality via the "social-construct precipitate," the central issue, then, is not so much whether relativity exists, but whether the way it is established and maintained is desirable; not whether universality exists, but whether the way it is being attained and maintained is desirable. These are not purely theoretical questions; they are socio-political questions as well. Hence, it is crucial that we keep in mind that power relations, though they take place at the level of the *presented*, affect minimally the *social given*. Power must be included in any discussion concerning the universality of humans. To underscore the effects that the exercise of power has had on the making of humanity in and through history as well as today is a task yet to be fully embraced by current social theory and research. It remains a gaping absence.

X

These considerations lead me to add the following note: In my view, it is not altogether clear that diffusiveness of cultural structures leads to greater freedom for the individuals concerned; on the contrary, it seems that diffusiveness may lead not to freedom but to exaggerated forms of individualism. In that case, individuals may come to see social forces at best as indifferent, alien at worst. The moral demands stemming from society would be too abstract to be meaningfully adopted. The dangers of that condition have been pointed out time and again by the founders of social theory: Marx's concern with alienation, which he retained, however transformed, throughout his life served him as a goading element in the search for the reconciliation of individual and society.[17] Durkheim's uncovering of *anomie* as the particularly modern social disease also prompted him to advocate sociology as the health science of the body social. His premature welcoming gesture to a society based on Organic Solidarity betrays the urgency with which he felt the need for a solution to the problem of *anomie*.[18] Freud's attempt to replace religion with science as the source of moral injunctions also points to the concern with a dis-eased society. Freud never advocated freedom from interdictions across the board to ease the tension that made neurotics out of citizens. On the contrary, if he sought to overthrow religion it was so as to better ground moral demands.[19] It was perhaps Weber who felt the daemons loose in the modern world, battling within his very being. He despaired. There is no vision of a good society in Weber, only a story of disenchantment and a veiled longing for an old authority. All these efforts are connected with the issue of the disintegration of society and the desire for overcoming that disintegration. That such a state of utter atomization may readily lead to its opposite, total integration, was most forcefully pointed out by Dostoyevski in his masterful portrayal of the Grand Inquisitor's political thought. Modern European history is a clear object-lesson.

It is in this context that the value of cultural diversity must be understood. Marx, although he was right about its import, was wrong in placing so many of his hopes in the destructive powers of capitalism to do away with cultural diversity and thus prepare the way for a more just society. For once all cultures have been destroyed, what would insure the effective force of moral demands? In a world of emancipated "humans," can there be meaningful individual fulfillment? The different cultures function very much as Durkheim and Hegel believed occupational groups ought to function in modern societies: they make cultural universals concrete and set affective, cognitive moral parameters that can be internalized by concrete individuals.[20] Without cultures, individual fulfillment is not possible. To speak in Freud's language, the introjection of excessively abstract "human demands" would give rise to an abstract and hence powerless Super Ego. Under such conditions individuals may become what Adorno thought he recognized already in his time: opportunists ready to sacrifice anything for survival. Cultural variety is not an anachronism, it is the most effective buffer against dehumanization.

Of course, what must be forgotten is that even when cultures do function as buffers against dehumanization, they need not be consciously constituted by all individuals concerned. To have lost sight of the power relations underlying the creation and maintenance of cultures was one of Durkheim's major shortcomings. How and why particular cultural values are created or reasserted must be open to reflection. This is the point behind Habermas' insistence on communication free from distortion. However, Habermas's own grounding of his theory seems to be a very Kantian, transcendental effort and hence subject to criticism for being itself too abstract.[21] But the animus behind Habermas' position needs to be retained. It may for instance be worthwhile considering Paolo Freire's discussion of *conscientizaçao*, the development of consciousness with a conscience, which suggests a deeper and closer connection between cultural values, self-reflection and practical life.[22] In this instance, as always, theory leads to practice. Whether the occasion for such practice can only take place at the periphery of modern society as Alasdair MacIntyre would want to claim, is a practical question.[23]

NOTES

* I thank Hugh Lacey, Joseph Margolis, Michael Krausz, Richard Eldridge, R.H. Wosniak, K. Ann Renninger, Robin Wagner-Pacifici and Harald Bershady for their comments.

1 The clearest discussion of radical positivism and its relation to social theory is to be found in Talcott Parsons' *The Structure of Social Action* (New York: Free Press, 1937).

2 Barry Barnes and David Bloor's discussion of relativism and social theory is only one of the latest examples of this misunderstanding of the sociological tradition. See Barnes and Bloor, "Relativism, Rationalism and the Sociology of Knowledge," in Steven Lukes and Martin Hollis, *Rationality and Relativism* (Cambridge: MIT Press, 1983).

3 Karl Marx, *Capital* (New York: International Publishers, 1967), p. 51.

4 Emile Durkheim, *The Division of Labor in Society* (New York: Free Press, 1964), p. 114.

5 Sigmund Freud, *Standard Edition* (London: Hogarth Press, 1953, 1974. Vol. XX), p. 198.
6 Sigmund Freud, *ibid*, Vol. XII, p. 115.
7 Max Weber, *Economy and Society*, Ed. Guenther Roth and Claus Wittich (New York: Bedminster Press), 1968. See especially Vol. 1, pp. 1–211.
8 Karl Marx, *Capital*, pp. 42–43.
9 This position is at the core of Durkheim's last major work, *The Elementary Form of the Religious Life* (New York: Free Press, 1965.
10 This position holds throughout Freud's work from the "Project for a Scientific Psychology" through Chapter VII of *The Interpretation of Dreams* to his later and more sociological writings. When Freud made the distinction between the primitive and the civilized, he meant to underscore precisely the similarity between the two.
11 Max Weber, *Economy and Society*, p. 5.
12 See Talcott Parsons, *Action and the Human Condition* (New York: Free Press, 1978), and *The Evolution of Societies* (Englewood Cliffs, Prentice-Hall, 1977).
13 The pertinent works by George Herbert Mead are *The Philosophy of the Present* (Chicago: Open Court Publishing Co., 1932); *The Philosophy of the Act* (Chicago: The University of Chicago Press, 1938); *Movements of Thought in the Nineteenth Century* (Chicago: University of Chicago Press, 1936); *Mind, Self and Society* (Chicago: University of Chicago Press, 1934).
14 This conception of social interaction was developed along Marxist lines in the Soviet Union by Lev Semenovich Vygotsky (1896–1934). Vygotsky seems to have been influenced by the work of John Dewey and William James, two of the leading members of the Pragmatist tradition in America and friends of G.H. Mead. Recently, J. Shotter has developed somewhat further this interactive view of social action in psychology. See L.S. Vygotsky, *Mind in Society: The Development of Higher Psychological Processes* (Cambridge: Harvard Univerity Press, 1978); J. Shotter and J. Newson, "An Ecological Approach to Cognitive Development: Implicate Orders, Joint Action and Intentionality," *Social Cognition: Studies in the Development of Understanding* ed. G. Butterworth and P. Light (Chicago: University of Chicago Press, 1982).
15 Richard J. Bernstein, *Beyond Objectivism and Relativism: Science, Hermeneutics and Praxis* (Philadelphia: University of Pennsylvania Press, 1983).
16 J. Fryed's work advances a similar position within psychology. See J.J. Fryed, "Sharability: the Social Psychology of Epistemology," *Cognitive Science* 7 (1983): 191–210.
17 Marx's communist society is nothing if it is not the proffered solution to the alienation of modern man and the social contradictions he detected undergirding the antinomies of bourgeois thought. But communism is not the harbinger of anarchy, nihilism or relativism. Perhaps this is most clearly seen in Marx's view of the nature of labor in the communist society. Labor would then be akin to artistic production in bourgeois society: it would be free but disciplines; it would constitute the purposive exertion of effort without the burden associated with alienated labor: "really free working, e.g. composing," wrote Marx, "is at the same time precisely the most damned seriousness, the most intense exertion." *The Grundrisse*, ed. and trans. David McLellan (New York: Harper & Row, 1971, pp. 611–612.
18 There is a fundamental contradiction in Durkheim's work. On the one hand, he asserts the emergence of an Organic Solidarity in modern society, that would integrate individuals more adequately within the body social. On the other hand, he notes that the most important and more extensive sector of modern society, the industrial/commercial sector, is shot through with *anomie* and de-regulation. He managed to understand such *anomie* as an aberration, a pathology. But this understanding was faulty by his own account: if normal phenomena are what are most widespread in any given society (*Rules*), how to explain the fact that what is most important and extensive in modern society is pathological (*Suicide*)? It is to Durkheim's credit that he did not, for the sake of consistency, attempt to undermine his own findings which showed modern society as pathological. Yet, to the end of his life he held on to the view that an Organic Solidarity was at hand. The good society, it would seem, would be born out of the pathological conditions. But without a Marxist reading of this state of affairs, Durkheim was left with only a hope and a contradiction. See Durkheim's *The Division of Labour in Society; The Rules of Sociological Method* (New York: Free Press, 1966); *Suicide* (New York: Free Press, 1966).

19 Discussing religion as an illusion, Freud notes: "If the achievement of religion in respect to man's happiness, susceptibility to culture and moral controls are not better than this, the question cannot but arise whether we do wisely in basing our cultural demands upon it." (*Standard Edition*, Vol. XXXI, p. 38). To Freud, religion is no longer an effective force and needs replacing with science. He follows a tradition which counts Feuerbach, Marx, and Nietzsche as some of its most notable representatives: God is a projection of humanity. In religions humans meet their "internal mental processes again outside themselves" (*Standard Edition*, Vol. XII, p. 92). For Freud the task is not to change social conditions so that religion fades out from history but to "change back once more by science into the *psychology of the unconscious*" these projected gods (*Standard Edition*, Vol. VI, p. 268). However, Freud understood science to be an austere maiden demanding, minimally, constant disciplined attention. It is echoing this understanding of science that Freud wrote in 1927 that *Logos* was his god-term (*Standard Edition*, Vol. XXI, p. 54); also, as he neared the end of his life, that the "dictatorship of reason" was our only hope. (*Standard Edition*, Vol. XXII, p. 213).

20 The idea of occupational groups is most forcefully presented in Durkheim's second preface to *The Division of Labour in Society* and in the lectures he delivered first at Bordeaux and later at La Sorbonne. See *Moral Education: A Study in the Theory and Application of the Sociology of Education* (New York: Free Press, 1961). Hegel's position on the same issue is clearly stated in *The Philosophy of Right* (London: George Bell and Sons, 1896).

21 For one of the latest readings of Habermas placing him squarely within the Kantian tradition, see Gillian Rose's excellent work, *Hegel Contra Sociology* (London: Athalone, 1981).

22 See Paolo Freire, *The Pedagogy of the Oppressed* (New York: The Continuum Publishing Corporation, 1970).

23 See Alasdair MacIntyre *After Virtue* (Notra Dame: University of Notre Dame Press, 1982).

RATIONALITY AND REALISM

JOSEPH MARGOLIS

In considering the question of rationality, one could hardly do better than begin with the splendidly helpful paper that Steven Lukes published some fifteen years ago – which assembles many of the most important alternative conceptions in the anthropological and sociological literature but which is nearly entirely wrong about the theoretical constraints on rationality itself;[1] or, turn to the quite opposed but genuinely searching attempt to relativize the question in accord with the influential program of the sociology of knowledge, offered recently by Barry Barnes and David Bloor – which, nevertheless, is utterly preposterous.[2] Luke's paper errs instructively, because, although it raises the essential issues (which it would be a misfortune to lose a grip on), it draws all the wrong conclusions by a series of *non sequiturs*. The reason, in part at least, undoubtedly lies with Lukes's strong opposition to all forms of cognitive relativism.[3] Barnes and Bloor's account tries to show the way to a balanced relativism, particularly with regard to rationality, but it bungles the effort in a double way: first, because none of the criteria the account offers for sorting relativisms do, singly or collectively, actually demarcate relativistic and nonrelativistic theories (or, more generously, do so for candidates that could really count as interesting versions of relativism); second, because the strategy of the adopted sociology of knowledge does not and cannot distinguish between the merely causal conditions of belief and the normative conditions for the validity of belief claims, or for the truth of beliefs, or for the possession of knowledge by believing agents – where the latter conditions positively entail a conception of rationality and the former merely presuppose such a conception (or, more desperately, suppose that we could work without one). Lukes's position, then, is relevant but quite wrong; and Barnes and Bloor's position, game but wildly irrelevant, even self-defeating. Of course, saying so is meant to signify that drawing the contrast is instructive about the question of rationality. To show this, however, requires a tolerance for what may seem a shift of topic.

Margolis, J., Krausz, M. and Burian, R.M., (eds.), Rationality, Relativism and the Human Sciences.
 ISBN 90-247-3271-9.

I

Lukes's thesis is quite straightforward. "Some criteria of rationality," he holds," are universal, i.e., relevantly applicable to all beliefs, in any context, while others are context-dependent, i.e., are to be discovered by investigating the context and are only relevantly applicable to beliefs in that context."[4] Lukes applies the distinction against Peter Winch's now nearly overly familiar discussion of Zande rationality and understanding.[5] But the way in which Lukes puts his point is obliquely instructive because it is so understandably hasty – without, however, vindicating Winch: "In so far as Winch seems to be saying that the answer to the . . . question [what for society *S* are the criteria of rationality *in general*?] is culture-dependent, he must be wrong, or at least we could never know if he were right; indeed we cannot even conceive what it would *be* for him to be right. In the first place, the existence of a common *reality* is a necessary precondition of our understanding *S*'s language . . .. What must be the case is that *S* must have our distinction between truth and falsity if we are to understand its language, for if *per impossibile* it did not, we would be unable even to agree about what counts as the successful identification of public (spatio-temporally located) objects. Moreover, any culture, scientific or not, which engages in successful prediction (and it is difficult to see how any society could survive which did not) must presuppose a given reality . . .. In the second place, *S*'s language must have operable logical rules and not all of these can be pure matters of convention . . .. It follows that if *S* has a language, it must, minimally, possess criteria of truth (as correspondence to reality) and logic, which we share with it and which simply *are* criteria of rationality . . .. But if the members of *S* really did not have our criteria of truth and logic, we would have no grounds for attributing to them language, thought or beliefs and would *a fortiori* be unable to make any statements about these."[6] What Lukes is opposing, here, is the thesis, attributed to Winch, that "rationality itself is context- or culture-dependent." Lukes sides with Alasdair MacIntyre, who holds that "we cannot explain social behavior independently of our own norms of rationality" *and* that "the beginning of an explanation of why certain criteria are taken to be rational in some societies is that they *are* rational"[7] – which in effect signifies that our understanding of the Azande, say, presupposes Lukes's own thesis about universal criteria of rationality.

Now all of this is quite mistaken – but nearly enough right to miss detection. Also, the required correction exposes the extraordinary complexity of the notion of rationality and with it the complexity of the status of the human sciences. These ramifications justify fussing over Lukes's and Barnes and Bloor's alternative accounts.

For example, Lukes claims that "the existence of a common reality is a necessary precondition of our understanding *S*'s language''; *and* that if that is so, then *S* and we must share the same ''distinction between truth and falsity,'' the same referential practice, the same ''operable logical rules'' (which coult not be ''pure matters of convention''), the same ''criteria of truth . . . and logic,'' in effect, the same ''criteria of rationality.'' Lukes never does explain what he means by the expression, ''our understanding *S*'s language''; or why, if our understanding presupposes a ''common reality,'' it follows that we must also share (among ourselves as well as between ourselves and the members of *S*) the same criteria of truth, reference, logic, rationality. The underlying *non sequitur* depends simply on *the presumed irreconcilability of realism and relativism*: on drawing the conclusion that *sharing the same world entails sharing the same conceptual scheme (of that common world) or the same cognitive criteria, or the same criteria of rationality*. It also supposes, too sanguinely, an easy resolution of questions regarding the status (pertinent to realist/anti-realist disputes) of theories, theoretical entities, the meaning of observationality, phenomenological laws, the entire apparatus of scientific method.[8] Ironically, the play of opposing views on just these questions actually encourages relativistic interpretations of both science and rationality.

Here, reference is often made to Donald Davidson's familiar account, in ''On the Very Idea of a Conceptual Scheme,''[9] to the effect that ''We cannot make good sense of the idea that there are seriously different total conceptual schemes, or frames of reference, or that there may be radically ''incommensurate'' languages (to use Whorf's word).''[10] What is usually not noticed, in Davidson's argument, is that *his* thesis depends (quite tellingly) on a familiar *reductio* of Whorf's hypothesis:[11] – if we can translate or interpret the verbal behavior of others, then our conceptual schemes cannot be ''radically incommensurate''; and if we can't, then there's nothing to be said about the properties of the language we fail to understand. The argument counts heavily against certain extravagances of T.S. Kuhn and Nelson Goodman – but it does have its limitation. In particular, when Davidson makes his claim, he intends ''conceptual scheme . . . to correspond to a whole language; nothing, at any rate [he says], can be left out that is needed to make sense of the rest.''[12] Fair enough. But conceding that has nothing to do with: (a) sharing all the same particular concepts, or (b) sharing any of the same criteria of truth, reference, logic, or rationality. It is quite possible, *contra* Lukes, to share a common world and, understanding one another, not to have ''radically incommensurate'' conceptual schemes, while diverging conceptually and while *not* sharing the very same cognitive criteria for getting around in our common world. In fact, Lukes himself gives the game away when – quite correctly – he notices that the view he espouses leads us to suppose that our jointly shared ''criteria of truth'' would in effect be *criteria* of ''correspondence to reality.'' *If*, however, all forms of foundationalism, essentialism, and the like are eschewed – as hopeless to vindicate[13] – then there *could* be no way to show that we must share the same *criteria* (of correspondence) in merely establishing our sharing a common reality. Alternatively put, there may be some use in a correspondence *theory* of truth, but (short, say, of subscribing to Russell's notion of knowledge by acquaintance) there is no use in correspondence criteria of truth – as opposed to *criteria* (of whatever kind) for assigning truth-values.

It is true – and important to emphasize – that the conception of rationality *is* a conception that cannot preclude an aptitude for applying cognitive criteria of some sort bearing on truth and reference and predication and the like; for not only must the beliefs, intentions, behavior of a set of creatures cohere in some suitable way but such creatures must be able to correct or adjust beliefs, intentions, behavior (in a suitable reflexive sense) on the basis of what, from their perspective, functions or counts as counterevidence. Lukes is quite correct to stress this connection.[14] It is just what links rationality and realism.

But it is *not* true that that conceptual linkage entailes that we share a "whole language" or "total conceptual schemes" (in Davidson's terms) – or (what is an entirely different issue) the same "criteria" of truth, reference, logic, or rationality (in Lukes's terms). And that means that, contrary to the contemporary fashion, *it is not the case that realism and a moderate (or robust) relativism are incompatible*. (Or at least, it need not be the case – since "relativism" is defined so arbitrarily or maliciously these days.[15]) In fact, it is important to notice: (i) that Davidson's argument against radically incommensurate conceptual schemes (and analogues of Davidson's argument – Max Black's, for instance) has, as such, nothing to do with the issue of realism, in any epistemically focused sense; (ii) that when they insist that, "For the relativist there is no sense attached to the idea that some standards or beliefs are really rational as distinct from merely locally accepted as such," . . . "that all beliefs are on a par with one another with respect to the causes of their credibility [or rationality] . . . [that is,] regardless of truth and falsity . . .,"[16] Barnes and Bloor deprive rationality of its necessary connection with normative functions and deprive relativism of any possible prospect of functioning as a rationally viable option; and (iii) Lukes's realism (minimal in terms of what it asserts, maximal in terms of what it infers from what it asserts) does not and cannot, contrary to what he supposes, yield any conclusions about universal conceptual schemes and universal criteria of truth, reference, logic, or rationality.

A short way of putting the point against Lukes is this: it is entirely possible that we share a common world but, for historically pertinent reasons, do not share the same paradigms or criteria for formulating and testing our cognitive claims. What this suggests is that to deny the possibility of that condition is tantamount to rejecting the historicity of human inquiry *and* to subscribing (once again) to foundationalism, essentialism, or some cognitively apt form of the correspondence theory. For *if*, as MacIntyre says, we must rely on "our own norms of rationality," *how* can we show that those very norms *are not* (in the terms of reproach Lukes favors against Winch) "culture-dependent," "context-dependent," historically local? At the very least, there is a *lacune* (the same one that can be traced in Charles Taylor, Hans-Georg Gadamer, and all those who admit the historical nature of human existence but deny even a moderate relativism).

II

There is a very instructive inadvertence in certain recent views of Hilary Putnam's, that can be made to lead us out of these difficulties. Putnam accepts an argument of Davidson's to the effect "that *truth is prior to reference* in the following way: a speaker's understanding of the reference of individual words consists in his grasping the truth conditions for sentences which contain those words . . .. Although formally reference appears to be prior to truth, in that logicians define truth in terms of reference, philosophically it is the other way around. What a speaker learns when he learns his native language is the truth conditions for its sentences, or for a subset of its sentences.[17] At least two points are worth pressing here. First, there *cannot* be any "priority" between words and sentences: words and sentences are linked in a conceptually relational way. It was in fact one of the fundamental mistakes of Quine's theory of language that he believed – and, evidently, Davidson and Putnam, following him – that truth conditions for sentences (Quine's notorious ersatz doctrine of "stimulus meaning") could be provided independently of *all* parsings in terms of constitutive words. But that's impossible: there cannot be any remotely *realist* reading of the truth of sentences without some notion, however inchoate, of reference to the world *via* the terms within such sentences and the terms of other sentences by means of which truth conditions can be confirmed or infirmed.[18] Furthermore, this sense of realism is entirely neutral to the issues at stake in recent disputes about realism and anti-realism.[19]

The issue is vexing when taken in Davidson's sense, which Putnam does not report in the fully relevant respect. For Davidson holds *only* that "an absolute definition of truth fails to yield an analysis of the concept of reference," where, by an "absolute" theory, one means a theory of truth "that satisfies something like Tarski's Convention T: . . . that by recursively characterizing a truth-predicate (say 'is true in L') entails, for each sentence *s* of L, a metalinguistic sentence got from the form '*s* is true in L if and only if *p*' when '*s*' is replaced by a canonical description of a sentence of L and '*p*' by a sentence of the metalanguage that gives the truth-conditions of the described sentence."[20] But an absolute definition is obviously *not* philosophically neutral with respect to truth-conditions as, on the contrary, is Tarski's account of the *meaning* of "true." It presupposes – what Davidson never actually shows is reasonable, and what cannot be deduced from the neutral Tarskian conception of "true" – the reducibility of the syntactic-semantic structure of natural languages to the canonical structures Tarski provided for selected formal languages. Here, Davidson specifically intends to constrast the "absolute" use of the primitive, undefined "truth-predicate" with any and all theories that "relativize truth to an interpretation, a model, a possible world, or a domain" – that is, with theories that assign "true" an *epistemic* force. The first achieves the result Davidson wants simply because "the accounts of truth and satisfaction [given in some language] don't suggest *how to go on to the new case* [when a new predicate or a new predicative application is added to the language in hand]."[21] True enough; but then Tarski's truth-predicate has no epistemic import at all (depends solely on formal equivalences) and his canonical program applies only doubtfully to whole natural languages – and cannot be applied by way of any merely formal or epistemically neutral procedure. It is the conflating of these two features of Tarski's work that gives the false impression of the power of Davidson's thesis. In any case, *reference*, in the epistemically live sense that Putnam favors,[22] *cannot* be posterior to truth along Davidson's lines: Davidson's accounts is epistemically irrelevant, and Putnam's cannot be construed as a merely formal notion.

The second point to press is simply that Putnam himself insists, in explicating what has been termed "the new theory of reference" (that he takes Kripke and himself to have developed – more or less against Russell's theory), that "the extension of certain kinds of terms ('natural kind words,' . . . meaning names for such things as natural substances, species, and physical magnitudes) is not fixed by a set of 'criteria' laid down in advance, but is, in part, *fixed by the world*. There are objective laws obeyed by multiple sclerosis, by gold, by horses, by electricity; and what it is rational to include in these classes will depend on what those laws turn out to be."[23] Hence, on Putnam's own theory, meaning and truth-conditions entail a rich concept of reference. Furthermore, Putnam indicates his own agreement with Michael Dummett's criticism of Davidson's account, to the effect that *two* distinct notions of truth are required: one, an "internal," purely formal "*equivalence principle*" (capturing Tarski's neutral truth-predicate), which has no explanatory power; the other, an "external," interpreted, justificatory notion that does have an "*explanatory role* in an account of understanding" – and which, in having that, indissolubly links truth and reference within a theory of meaning.[24]

Putnam disagrees with Dummett's apparent collapsing of truth into justification, of course – essentially on the grounds that truth but not justification "cannot be lost," is not tensed or relativized or a matter of degree.

Hence, Putnam holds: "truth [is] an *idealization* of justification" – which marks his well-known contrast between "internal realism" and "metaphysical realism."[25] But the fact remains that, on Putnam's view, the meaning (hence, the truth-conditions for the use) of natural kind terms (which, presumably, organize our sciences) is not conventional or subjective, and positively requires determinate *reference* to the real world (at least reference as determinate as predication itself). Furthermore, on Putnam's view, the rationality of human behavior can apparently be gauged by behavior and dispositions favorable or unfavorable to bringing our terms into line with these realist considerations. So truth (*a fortiori*, meaning) and reference are conceptually symbiotic.

Once we accept this much, it is but a step to see as well that – on any "interpreted" theory of truth that resists doctrines functionally similar (in any of a very large number of ways) to (say) Russell's theory of knowledge by acquaintance (to what in a larger sense is "foundationalism" or, in the Continental idiom, the "philosophy of presence" or "logocentrism") – *there is and can be no principled basis for insuring uniquely determinate truths about the world*. Perhaps the point should be put in a somewhat different way. By *realism* – in an epistemically fundamental sense – we may mean either of two doctrines: (i) there is an actual world, structured in some way independent of human inquiry, that human inquirers have (some) knowledge of; (ii) there are denumerably many facts about the actual world as it is structured independent of human inquiry, that human inquirers have (some) knowledge of. (i) is compatible with supporting a realism (*sans phrase*) in which the ingredients of our science, sorted in terms of the structures of the world independent of inquiry and the structures of inquiry that affect our access to the independent world, *cannot* be disjoined from one another (roughly, the upshot of Kant's view of the relationship between transcendental idealism and empirical realism, without any concession to Kant's own theory of transcendental reasoning or of constraints on empirical knowledge); (ii) entails a disjunction between such "realist" and "idealist" ingredients (roughly, the upshot of admitting that Putnam's contrast between "internal" and "metaphysical" realism does specify – against Putnam's own intention – viable alternatives,[26] without any further consideration of Putnam's own theories of language or of constraints on empirical knowledge).

Broadly speaking, (i) requires that realism be an *en bloc* doctrine – what may fairly be termed a *pragmatist* doctrine, in the sense that though there may be a (second-order) ecological or evolutionary or survivalist justification for realism, there cannot be a (first-order) cognitivist justification, a particular cognitive discrimination that confirms the fact of the matter. This is also why there *cannot* be a first-order testable *bloc* science, in Quine's sense, though there can be a first-order *bloc* testing of "a whole group of hypotheses" within the larger whole of science, in Duhem's sense.[27] To see matters this way is, in effect, to hold that, if we subscribe to a realism in accord with (i), we cannot, merely on those grounds, draw Lukes's conclusions about sharing common concepts or common criteria of truth, reference, logic, or rationality. Lukes, in effect, has suppressed his conviction that the only viable realism – hence, the only convincing ground for a viable form of rationality, taken as having an epistemic function – *must* be a form of (ii) metaphysical realism, or a realism that makes no concessions to idealism (that is, a kind of foundationalism or essentialism). But that, of course, is both indefensible in itself and (*a fortiori*) significantly harder to defend than any version of (i).

It is a further – stronger – thesis, going very far beyond merely demonstrating the *non sequitur* of Lukes's argument, that, given a realism in accord with (i), *and* the denial of (ii), there are indefinitely many (shall we say) "cognitively defensible" accounts of the constitutive phenomena of the world, if there is any such account. Putnam hesitates here between characterizing this position – in effect, his own – as internal realism or internal nonrealism, because of his wish to reconcile the executive metaphysical theories of Quine, Goodman, and Dummett at least. He offers the following useful distinction: "On [the] perspective [of metaphysical realism], the world consists of some fixed totality of mind-independent objects. There is exactly one true and complete description of 'the way the world is.' Truth involves some sort of correspondence relation between words and thought-signs and external things and sets of things . . .. [On] the *internalist* perspective, . . . *what objects does the world consist of*? is a question that it only makes sense to ask *within* a theory or description. Many 'internalist' philosophers, though not all, hold further that there is more than one 'true' theory or description of the world. 'Truth,' in an internalist view, is some sort of (idealized) rational acceptability – some sort of ideal coherence of our beliefs with each other and with our experiences *as those experiences are themselves represented in our belief system* – and not correspondence with mind-independent or discourse-independent 'states of affairs.'"[28] Putnam also wishes to contrast the internalist view with the relativist's, but that's because (prejudicially) he simply defines these alternatives in such a way that: "the non-realist position [the internalist's] . . . assumes an *objective* notion of rational acceptability"; whereas, although it also rejects metaphysical realism, the relativist's regards "truth or rational acceptability as *subjective*": "The whole *purpose* of relativism, its very defining characteristic [he says], is . . . to *deny* the existence of any intelligible notion of *objective* 'fit'. Thus the relativist cannot understand talk about truth in terms of *objective* justification-conditions."[29]

Two distinctions should put everything into manageable order. First of all, Putnam contrasts internalism and relativism in the way he does (understandably but arbitrarily), because he takes Kuhn and Feyerabend – as champions of the incommensurability thesis – to be paradigmatic relativists (hence, also, subjectivists);[30] but it is entirely possible (and entirely reasonable) to formulate relativism as not Protagorean or skeptical or subjectivist or anarchist, but committed instead (without contradiction) *to* cognitively viable notions of "objective fit," *within the constraints of an internalist metaphysics*. So seen, Putnam may be said not to have shown that (or even how) the internal realist (or nonrealist) can fail to be a relativist or at least to accommodate the relativist (who could easily be one of Putnam's own favored sort of theorist, that is, one committed to the notion of objecttive fit, consistently with affirming "more that one 'true' theory or description of the world" not mutually compatible or reconcilable with one another). Secondly, if "objective fit" is relativized to conceptual scheme (though not in the radical sense Davidson opposes), and if no metaphysical realism can force or lead us to the "one true and complete description of 'the way the world is'" – which is certainly not tantamount to incommensurability – then, just as *truth claims* or *truth values* (as opposed to truth *tout court*) are relativized to conceptual schemes, *so too is reference*. "'Objects' do not exist independently of conceptual schemes [says Putnam] . . .. Since the objects *and* the signs are alike *internal* to the scheme of description, it is possible even trivial to say what matches what."[31] "Trivial" is actually not a helpful charge here (though the point is clear enough), once we treat linguistic usage in a richly historicized way, pluralized for different societies, and specialized in terms of a division of labor. The deeper insight remains that truth, meaning, and reference are conceptually inseparable distinctions for any realist conception of science compatible with (i) above, whether or not one approaches matters from the view of metaphysical or internal realism and whether or not one is a relativist. Furthermore, *if* (and *because*) rationality is a notion that has an epistemic function to play – linking human behavior to our very mode of understanding reality – *the theory of what it is to be rational cannot be prised apart from the theory of science*.

A final word of warning is in order. In effect, Putnam *is* a kind of moderate conceptual relativist, provided only that (on his view) alternative "internalist" characterizations of the world preserve a measure of "objective fit" *and* that alternative conceptual schemes are not attenuated so far that they exhibit the extreme incommensurability he opposes in Kuhn and Feyerabend. Nevertheless, quite ironically, he is himself tempted by a more extreme form of relativism than the moderate relativist is likely to favor. This is undoubtedly due to his attraction to Quine's thesis about the "inscrutability of reference" and Davidson's thesis that "truth is prior to reference."[32] But it also explicitly involves his appeal to the Skolem-Löwenheim theorem.[33] From purely formal considerations of that theorem – which of course have no direct connection with the actual conditions of cognitive inquiry of an empirical sort – Putnam is led to affirm that "the truth conditions for 'A cat is on a mat' don't even exclude the possibility that 'cat' refers to *cherries*."[34] But, from the modestly relativistic concession that reference cannot be uniquely fixed under the conceptual conditions acknowledged, it hardly follows that there is any radical indeterminacy or inscrutability of reference that *would "fit" the actual conditions of human existence and human inquiry*. Furthermore, *if* predication would not support a comparably radical finding, and *if* predication and reference *are* conceptually linked in epistemic contexts, then the radical view Putnam espouses is both untenable and tantamount to utter conceptual chaos. The elementary point is that the Skolem-Löwenheim theorem is entirely formal and *lacks altogether any linkage with human cognitive constraints pertinent to a realist science* – realist in at least the minimal sense of distinction (i), offered above.

III

This leads us back to Lukes's sanguine *non sequitur* and the heroically self-defeating irrelevance of Barnes and Bloor's reductive (putatively relativistic) sociology of knowledge. For, if only a form of internal realism is viable (or internal nonrealism, or anti-realism or even irrealism – to accommodate, without prejudice, here, all candidate theories that are strongly pragmatist, opposed to metaphysical realism, and committed to the symbiosis of realism and idealism), then the shift to internal realism is bound to affect substantively, decisively, possibly even radically, what we may defend as an adequate theory of rationality.[35] For example, if we adopt a strongly historicized and praxicalized account of concepts and human understanding – which Lukes appears to favor[36] – then it is quite impossible to reach any of the universalized conclusions regarding rationality Lukes himself draws *from* a minimally realist thesis. This is not to say that cultural or rational universals are impossible to confirm empirically. It is to say rather that *universalism* is incompatible with historicism, with praxical accounts of understanding, with pragmatism, with internal realism deprived of foundationalist pretensions (in effect, with kantianism); and that would-be *universals* are themselves only *general* uniformaties offered as candidates for universality *via* critical inferences to the best explanation of human life.

Alternatively put, social understanding, socially shared identification of objects, and the like, do *not* presuppose species-wide universals or species-wide criteria of truth and rationality: an empirically accessible conservatism of a conceptual and cognitive sort (undoubtedly linked to biological uniformities and to the relatively slow pace with which biologically constrained conceptual variability occurs within different societies and within the same societies diachronically considered) is quite sufficient to account for the *contingent* stability of such understanding. There is no other recourse. *The fact is that, under the constraints conceded, no one can successfully demonstrate what timeless conceptual uniformities obtain, that account for obviously effective social understanding.*

Historicism and universalism are simply incompatible; but universals themselves are empirically contingent general regularities sorted – under historical or praxical constraints – as the likeliest universalizable candidates we have. This is what, ultimately, is missed in those last, desperate versions of would-be timeless (totalized and trans-historical) theories of rationality advanced by Jürgen Habermas and Karl-Otto Apel.[37] Lukes cannot draw the conclusion he wants – from his own assumptions; and Habermas and Apel actually violate the conceptual constraints under which they attempt to theorize about rationality.

Carl Hempel provides the counterpart universalist conception of rationality favored by the partisans of the unity of science program, that is, by those largely indifferent to historical and praxical contingencies.[38] But that alternative is unsatisfatory for counterpart reasons: first, because it is doubtful that, if human psychology and the disciplines that depend upon the phenomena it examines cannot be reduced physicalistically – which seems most plausible[39] – then it will be impossible in principle to match nomic universals with all bona fide instances of causal interaction; second, because we have no clear strategy for insuring the universalism of the hypothetico-deductive model of explanation, once empirical generalizations of the causal sort are denied approximative status with respect to universal covering laws – for instance, by attacks on essentialism, by attention to ineliminable referential contingencies, by attention to the absence of empirical evidence exhibiting probabilized regularities as (invariably) tending asymptotically toward fixed limits of variation, and by the straightforward demonstation that, if interaction be conceded, then even the domain of physical forces (certainly that portion intimately involved in human action) cannot form a closed domain. Here, we may recall Davidson's rather gentle criticism of Hempel's account of explaining rational action under covering laws: "in the sense in which rationality is a trait that comes and goes, it can't be an assumption needed for every reason explanation. People who don't have the trait are still agents, have reasons and motives, and act on them."[40] But further, Davidson's attenuated acceptance of Hempel's nomological approach cannot but be similarly undermined by the requirement (his own) of "strict laws [for the domain in question: psychology] belonging to a *closed* system of laws and ways of describing events [where] there are no such laws governing the occurrence of events describing in psychological terms."[41]

We are being forced inexorably, by these and similar considerations, to concede that the viability of candidate theories of rationality and the very structure of such theories cannot be less fundamentally affected by historicized and praxical constraints than the theory of science itself. Once we give up metaphysical realism (in all its protean forms), reductive physicalism (in all its forms), the "closed system" status of the domains of the human sciences, the question of rationality will be seen to be more of a puzzle – of a piece with the theory of science itself – than an independent source of assured universal constraints (of the sort Lukes obviously favors) by means of which to temper our account of social understanding and the methodology of science. The intelligibility of human behavior, infra- and inter-culturally or infra- and inter-linguistically, will, then, be seen to be the result (in some sense) of the joint functioning of general biological uniformities (most plausibly linked to cognitive aptitude) *and* the radically historicized, contingent hermeneutic aptitudes of particular human beings.

Quite sensibly, Hempel (and with him, Davidson) took the hypothetico-deductive model of explanation in science to provide the most promising model – or, a salient part of that model – of rationality: a rational agent, on that view, is one who behaves (implicitly or explicitly) in accord with the most powerful cognitive model we possess. The generally admitted fact of scientific progress draws us in this direction. The trouble is that that model is in the deepest trouble itself, because the logical status of scientific laws is in doubt, because the deductive model of explanation is in doubt, and because the conceptual connection between explanatory models and the living, historical practice of scientific inquiry is unclear. This, of course, is precisely what is associated with the difficulties unearthed by such investigators as Kuhn, Lakatos, and Feyerabend.[42] We simply do not have a suitable model of the rationality of science, though we have some clues.[43] Consequently, when theorists construct a rational model of technology and purposive behavior as guided by the universal laws of nature – Mario Bunge, perhaps most notably[44] – they cannot fail to be open to the same difficulties that inhere in our attempts to model the rationality of science.

IV

Amid all these difficulties, however, one conceptual clue stands out above all others: we must concede that man is rational even if we are unable to model his rationality convincingly. We cannot merely *assume* human rationality – because of the reflexive nature of the would-be hypothesis. At present, man is the sole paradigm of the scientist, the cognitive agent, the psychologically endowed creature, the language user, the purposive and rational being. The study of all else (the *human* study of all else) is conceptually dependent on this condition[45] – which is at once the joint theme of analytic forms of realism and anti-realism and of the hermeneutic dependence of the physical on the human sciences. But the structure of all the roles mentioned is hardly self-evident: rationality itself is a second-order, contingently persuasive ascription – within historicized reflection – of the stablest general features of our own *cognitively* focused existence, that seem to stand the best chance of universally fitting the achievements of science and intelligent behavior. There is, inevitably, a normative function to be assigned science and intelligence and, correspondingly, a equilibrating congruity between these and rationality itself. This is the sense in which rationality is ultimately construed in pragmatist terms: it marks the effective survival of the human species *via* whatever its investigative and testing practices may be; it marks it, that is, *en bloc, not* in particular privileged propositions.

Hence, Barnes and Bloor are quite right to discredit Lukes's and Martin Hollis's thesis "that all cultures share a common core of beliefs and rationally-justified patterns of inference," "cultural universals" that insure (against what, in an impoverished sense, they themselves take to be relativism) the sharing of truth and the mutual understanding of men of different cultures.[46] They rightly remark: "reality is, after all, a common factor in all the vastly different cognitive responses that men produce to it. Being a common factor it is not a promising candidate to field as an explanation of that variation."[47]

Nevertheless, they fail to grasp the puzzle of rationality, because they impoverish both relativism *and* the conditions of cognition. Regarding the first, they offer three would-be features: (i) "that beliefs on a certain topic vary" – which cannot differentiate between relativistic and nonrelativistic theories; (ii) "that which of these beliefs is found in a given context depends on, or is relative to, the circumstances of the users" – which again is entirely neutral; and (iii) – their so-called "equivalence postulate" – "that all beliefs are on a par with one another with respect to the causes of their credibility. It is not that all beliefs are equally true or equally false, but that regardless of truth and falsity the fact of their credibility is to be seen as equally problematic."[48] Now, part of the point of (i)–(iii) is to disallow all versions of metaphysical realism, of foundationalism, of what Hollis terms an epistemically neutral "bridgehead" linking all culturally variable forms of life.[49] But part of the point Barnes and Bloor wish to support is that relativism (on their reading) is tantamount to an utter skepticism with respect to a rationally justified confidence in *any* form of reasoning; so they claim, explicating (iii), that "Our two basic modes of reasoning [deduction and induction] are in an equally *hopeless* state with regard to their rational justification."[50] Rationality, then, cannot be linked (as with Lukes) to any cognitively assured set of beliefs or any timelessly secure form of reasoning. Fair enough. But that hardly shows that relativism is equivalent to Barnes and Bloor's own extreme form of skepticism, or that the rejection of theories of rationality in accord with metaphysical realism leads directly to skepticism, or that the retreat to the sociology of knowledge (on their reading: the sociologist must eschew evaluating beliefs as true or false and as rational or irrational because of links to truth and falsity, and "must search [instead] for the causes of [their] credibility),[51] *or* (most damagingly) that it is even possible to explore the causes of beliefs *without a positive commitment to cognitive appraisal and to some positive view of rationality*. In this sense, Barnes and Bloor's program is utterly self-defeating.[52]

The required conception of rationality, then, must situate us between the extremes of metaphysical realism and mere skepticism; it must be cognitively sanguine, hence realist; it must, in escaping universalism, essentialism and the like, accommodate the profoundly historicized and praxical nature of human existence; *and*, as a second-order thesis, it cannot, on pain of contradiction, be formulated in *cognitivist* terms. If these constraints be conceded, then there is only one – extremely persuasive – option before us: some version of Charles Sanders Peirce's biologically motivated view of human rationality (suitably) spare, suitably adjusted, suitably relieved of Peirce's own fangango, neutral to other controversies) must be at least as close to the *beginning* of a satisfactory orientation as we could possibly get. A few tantalizing lines will confirm this: "every simple truth of science is due to the affinity of the human soul to the soul of the universe, imperfect as that affinity no doubt is";[53] "Certain general ideas of action [that is, the regularities of natural forces], prevail throughout the universe, and the reasoning mind is itself a product of the universe. These same laws are thus, by logical necessity, incorporated in his own being."[54] The point of Peirce's floridly expressed conception is that rationality cannot be an explicitly formulizable procedure, that the "abductive" or hypothesis-forming ability that we have cannot be justified on the grounds of rational strategies that are more than heuristic devices or recipes (against, say, Mill's Canons), that without the "natural prompting" that belongs to our having successfully evolved within nature itself, if we had "to search blindfold for a law which would suit the phenomena, our chance of finding it would be as one to infinity."[55] This means that we *are* justified in formulating would-be canons of reasoning, that there must be promisingly stable but changing methodological regularities (not assured a timeless validity) that our evolutionary and ecological achievement insures in on empirically contingent way. In short, realism, whether of the externalist or internalist sort, is unintelligible without a counterpart theory of human rationality.

Rationality, then, is a theory of an explanatory sort that we know no way of avoiding – in attempting to understand, reflexively, our own cognitive aptitude. That we have a cognitive aptitude is, broadly speaking, a biological fact, in the sense that the survival of our species depends on the intervention of sustained inquiry. It is a theory of a cognitivist sort *if* we subscribe to some form of metaphysical realism (or a cognate doctrine); and it cannot be cognitivist *if* we subscribe to a realism of the non-"metaphysical" sort here supported, which, otherwise (paired in Putnam's way, in terms of internal and external realisms), would remain seriously ambiguous. So rationality is, minimally, a species-specific model of the cognitively informed capacity of creatures – human and non-human – to survive, in the biological sense of reproducing populations. Cognition is justified on noncognitivist grounds, then, which is why rationality must in large part be accounted for in terms of "tacit" congruities between the uniformities of nature and the uniformities of mind.

Of course, there is also the familiar temptation to construe the rational as the moral or in some such way – that is, as providing normative direction for human conduct and behavior. In the same sense in which there must be general uniformities that yield recipes for scientific inquiry, so must there also be general prudential concerns that any rationally "responsible" (*species-wide*) policy must accommodate.[56] But once we view matters thus, we must see that any *infra*-species-wide model of rationality, in particular, normative action-guiding models for particular societies or individuals, cannot fail to be ideologically partisan or tolerant of a plurality of non-converging partisan models – historically situated, context-bound, to use Lukes's own terms (applied, ironically, almost as terms of opprobrium) At any rate, such models cannot be expected to draw supprt from those models that (in the internal-realist sense) minimally fit the admitted achievement of human science, even though that achievement must, rationally, also inform and constrain intelligent behavior. The attempt to draw universal ethical norms from inquiry itself was of course dear to Peirce (his "logical socialism") and to such philosophers as Habermas and Apel, who have tried to improve on Peirce in this respect.[57] But apart from the inevitable failure of their efforts, which follows from the incompatibility of historicism and universalism, any attempt to merge rationality and morality (or any doctrine of normative policy) must enrich our realism well beyond anything that the *explanatory* function of rationality would require or could support.

Rationality in the latter sense, precisely because it trades on biologically subcognitive uniformities in the species, in virtue of which we suppose man to have an aptitude for discerning selected uniformities in nature *and* for cognitively accommodating the conceptual orientation of diverse societies, *does not entail* (*contra* Lukes) universal criteria of rationality, universal criteria of truth, universally shared concepts. All that is required is that given criteria, given concepts, given beliefs, given theories are and can be shared by *particular* aggregates of cognitively apt agents actually in communication with one another: moving from one such group to another, there is no reason to suppose that universal and invariant concepts and criteria are employed in each. The success of *particular* communications does not require it – which of course goes against Apel and Habermas. Endorsing the search for normative universals that would fit actual communicative success concedes the explanatory irrelevance of such a search, as well as the futility of supposing that there could be only one such thesis suited to, or ideally convergent or congruent with, all candidate generalities of the pertinent sort. If, in spite of being realists in the minimal sense provided, we cannot insure a uniquely correct scientific account of the world – in a sense, the point of convergence between Kuhn and Putnam – then we cannot insure a uniquely correct account of rationality either.[58] Realism and rationality are conceptually linked concepts; but their connection is fully compatible with, and, given current views of realism itself, rather economically and favorably served by, a modest or robust relativism.

NOTES

1 Steven Lukes, "Some Problems about Rationality," *Archives Européennes de Sociologie*, VIII (1967).
2 Barry Barnes and David Bloor, "Relativism, Rationalism and the Sociology of Knowledge," in Martin Hollis and Steven Lukes (eds.), *Rationality and Relativism* (Cambridge: MIT Press, 1982).
3 Cf. "Introduction" in Hollis and Lukes, *op. cit.*; and Steven Lukes, "Relativism: Cognitive and Moral," *Proceedings of the Aristotelian Society*, XLVIII (1974). Lukes is somewhat drawn – ultimately opposed – to moral relativism; cf. pp. 171–172 and sec. iv.
4 Lukes, "Some Problems about Rationality," 260. This pretty much recapitulates "Relativism: Cognitive and Moral."
5 Peter Winch, "Understanding a Primitive Society," *American Philosophical Quarterly*, I (1964).
6 *Op. cit.*, 260–262.
7 Alasdair MacIntyre, "A Mistake about Causality in Social Science," in P. Laslett and W.G. Runciman (eds.), *Philosophy, Politics and Society*. Second Series (Oxford: Basil Blackwell, 1962), p. 61. The reference is cited by Lukes and the thesis is opposed by Winch.
8 See for example Hilary Putnam, *Meaning and the Moral Sciences* (London: Routledge and Kegan Paul, 1978); Bas van Fraassen, *The Scientific Image* (Oxford: Clarendon, 1980); Nancy Cartwright, *How the Laws of Physics Lie* (Oxford: Clarendon, 1983); and Ian Hacking, *Representing and Intervening* (Cambridge: Cambridge University Press, 1982).
9 Donald Davidson, "On the Very Idea of a Conceptual Scheme," *Proceedings and Addresses of the American Philosophical Association*, LXVII (1973–74).
10 Donald Davidson, "Psychology as Philosophy" (Comments and Replies), *Essays on Actions and Events* (Oxford: Clarendon, 1980), p. 243.
11 See Max Black, "Linguistic Relativity: The Views of Benjamin Lee Whorf," *Models and Metaphors* (Ithaca: Cornell University Press, 1962).
12 "Psychology as Philosophy," p. 243. Cf. Charles Taylor, "Rationality," in Hollis and Lukes, *Rationality and Relativism*.
13 For purposes of economy, we may take it that Richard Rorty has summarized the principal analytical arguments against versions of foundationalism, in *Philosophy and the Mirror of Nature* (Princeton: Princeton University Press, 1979). See also, Nicholas Rescher, *The Coherence Theory of Truth* (Oxford: Clarendon, 1973).
14 The point is not really featured by Davidson, who speaks rather more of testing rationality than

of its conceptual connection with being a cognitively apt and critical agent: it's not enough to have beliefs, or beliefs that guide action, or beliefs that guide action successfully. See Donald Davidson, "Hempel on explaining Action" and "Mental Events," in *Essays on Actions and Events*. The point is central, however, in Jaakko Hintikka, *Knowledge and Belief* (Ithaca: Cornell University Press, 1962).

15 See Hilary Putnam, "Beyond Historicism," *Philosophical Papers*, Vol. 3 (Cambridge: Cambridge University Press, 1983).

16 *Op. cit.*, pp. 27, 23.

17 *Op. cit.*, 291–292.

18 See W.V. Quine, *Word and Object* (Cambridge: MIT Press, 1960), Ch. 2. Certainly, one of the most celebrated remarks of the book is the following: "Occasion sentences and stimulus meaning are general coin; terms and reference are local to our conceptual scheme," p. 53. See also, Joseph Margolis, "Behaviorism and Alien Languages," *Philosophia*, III (1973).

19 See Hilary Putnam, *Philisophical Papers*, Vol. 3.

20 Donald Davidson, "Reality without Reference," in Mark Platts (ed.), *Reference, Truth and Reality* (London: Routledge and Kegan Paul, 1980), pp. 133, 131.

21 *Ibid.*, pp. 131, 133.

22 "Beyond Historicism," pp. 295–296.

23 Hilary Putnam, "Reference and Truth," *Philosophical Papers*, Vol. 3, p. 71.

24 *Ibid.*, pp. 82–83. Dummett's William James Lectures, which apparently develop these issues, have yet to be published. See also, Michael Dummett, "What is a theory of meaning? (II)", in Gareth Evans and John McDowell (eds.), *Truth and Meaning* (Oxford: Clarendon, 1976), p. 74.

25 *Ibid.*, pp. 84–86. See also, Hilary Putnam, *Reason, Truth and History* (Cambridge: Cambridge University Press, 1981), Ch. 3.

26 Hilary Putnam, "Realism and Reason," *Meaning and the Moral Sciences*; "Two Philosophical Perspectives," *Reason, Truth and History*: "Why there isn't a ready-made world," *Philosophical Papers*, Vol. 3.

27 See Pierre Duhem, *The Aims and Structure of Physical Theory*, trans. Philip Wiener (Princeton: Princeton University Press, 1954), Ch. 6; W.V. Quine, "Two Dogmas of Empiricism," *From a Logical Point of View* (Cambridge: Harvard University Press, 1953). There is a useful compendium of the principal views about the so-called Duhem-Quine thesis, in Sandra G. Harding (ed.), *Can Theories Be Refuted?* (Dordrecht: D. Reidel, 1976).

28 "Two Philosophical Perspectives," *Reason, Truth and History*, pp. 49–50.

29 "Two Conceptions of Rationality," *Reason, Truth and History*, p. 123.

30 *Ibid.*, pp. 113–119.

31 "Two Philosophical Perspectives," p. 52. Cf. also, "Models and Reality," *Philosophical Papers*, Vol. 3.

32 Cf. Quine, "Ontological Relativity," *Ontological Relativity and Other Essays* (New York: Columbia University Press, 1969).

33 See Hilary Putnam, *Reason, Truth and History*, Ch. 2 and Appendix.

34 *Ibid.*, p. 35.

35 The question about metaphysical realism is complicated in a further regard by certain pertinent remarks of Michael Dummett's. See Michael (M.A.E.) Dummett, "What is a theory of meaning?" in Samual Guttenplan (ed.), *Mind and Language* (Oxford: Clarendon, 1975); and "What is a theory of meaning? (II)."

36 Cf. Steven Lukes, *Power: A Radical View* (London: MacMillan, 1974), Ch. 4.

37 Jürgen Habermas, *Communication and the Evolution of Society*, trans. Thomas McCarthy (Boston: Beacon Press, 1979); Karl-Otto Apel, *Toward a Transformation of Philosophy*, trans. Glyn Adey and David Frisby (London: Routledge and Kegan Paul, 1980). Thus seen, John Searle's insistence on the necessary and sufficient conditions of speech acts is, apart from its departure from John Austin's original orientation, a pale version of Habermas's argument, in spite of the fact that Habermas obviously relies on the prior validity of Searle's and similar accounts. Cf. John Searle, *Speech Acts* (Cambridge: Cambridge University Press, 1969).

38 Carl G. Hempel, "Rational Action," *Proceedings and Addresses of the American Philosophical Association*, Vol. XXXV (1962).

39 See Joseph Margolis, *Philosophy of Psychology* (Englewood Cliffs, N.J.: Prentice-Hall, 1984); *Culture and Cultural Entities* (Dordrecht: D. Reidel, 1983); *Persons and Minds* (Dordrecht: D. Reidel, 1978).

40 Donald Davidson, "Hempel on Explaining Action," *Essays on Actions and Events*, p. 266.

41 *Ibid.*, p. 262; italics added. Cf. "Mental Events," *Essays on Actions and Events*; and Joseph Margolis, "Prospects for an Extensionalist Psychology of Action," *Journal for the Theory of Social Behavior*, II (1981).

42 The full adequacy of their particular views is not at stake.

43 See for instance, Thomas S. Kuhn, "Second Thoughts on Paradigms," *The Essential Tension* (Chicago: University of Chicago Press, 1977); Ernan McMullin, "Values in Science," *Proceedings, Philosophy of Science Association*, Vol. 2 (1982) – I have seen this essay only in manuscript; W.H. Newton-Smith, *The Rationality of Science* (London: Routledge and Kegan Paul, 1981); Kurt Hübner, *Critique of Scientific Reason*, trans. Paul R. Dixon, Jr. and Hollis M. Dixon (Chicago: University of Chicago Press, 1983). How remarkably scanty our clues here are may be judged from these discussions.

44 Mario Bunge, "Toward a Philosophy of Technology," in Carl Mitcham and Robert Mackey (eds.), *Philosophy and Technology* (New York: Free Press, 1983). Cf. Joseph Margolis, "Three Conceptions of Technology: Satanic, Titanic, Human," in Carl Mitcham (ed.), *Philosophy and Technology*, Vol. 7 (Greenwich, Conn. JAI Press, forthcoming). Of course, it is Imre Lakatos who is the principal (failed) advocate of a discovery of a model of rationality drawn from the historically shifting forms of scientific inquiry but not prejudicially linked to merely favoring any particular contingency of such inquiry. Cf. Paul Feyerabend, *Against Method* (London: NLB, 1975), Ch. 17.

45 See "Animal and Human Minds," *Culture and Cultural Entities.*

46 *Op. cit.*, p. 35. Cf. also, Martin Hollis, "The Social Destruction of Reality," in Hollis and Lukes (eds.), *Rationality and Relativism.*

47 *Ibid.*, p. 34.

48 *Ibid.*, pp. 22–23.

49 Cf. Hollis, "The Social Destruction of Reality."

50 *Op. cit.*, p. 41; italics added.

51 *Ibid.*, p. 23.

52 A discussion quite sympathetic with the view here pressed may be found in Russell Keat and John Urry, *Social Theory as Science* (London: Routledge and Kegan Paul, 1975), pp. 204–212.

53 *Collected Papers of Charles Sanders Peirce*, ed. Charles Hartshorne and Paul Weiss (Cambridge: Harvard University Press, 1931–1935), 5.47.

54 *Ibid.*, 5.603.

55 *Ibid.*, 6.10. Cf. Michael Polanyi, *Personal Knowledge*, corr. ed. (Chicago: University of Chicago Press, 1962).

56 I have tried to explore such minima in *Negativities. The Limits of Life* (Columbus: Charles Merrill, 1975). For a recent sample of action-guiding conceptions of rationality, largely focused in individualistic terms, see *Social Theory and Practice*, IX (1983): special double issue, "Rational Actions," ed. R.B. Brandt.

57 Karl-Otto Apel, *Charles S. Peirce: From Pragmatism to Pragmaticism*, trans. John Michael Krois (Amherst: University of Massachusetts Press, 1981).

58 For a compendious account of the current state of rationality theory, see Jon Elster, "Rationality," in Guttorm Fløistad (ed.), *Contemporary Philosophy*, Vol. 2: *Philosophy of Science* (The Hague: Martinus Nijhoff, 1982). Elster may well equivocate on "rationality" – as a theory of action within which "rational" and "irrational" behavior may be sorted, or as a theory of the normatively preferable alternative itself (cf. p. 120). For he appears to favor the thesis "that intentionality does not presuppose rationality," which suits neither option. Cf. also, Charles Taylor, "Interpretation and the Sciences of Man," in Paul Rabinow and William M. Sullivan (eds.), *Interpretive Social Science: A Reader* (Berkeley: University of California Press, 1979).

INDEX OF NAMES